1987

The Three Mile Island Accident

ACS SYMPOSIUM SERIES **293**

The Three Mile Island Accident
Diagnosis and Prognosis

L. M. Toth, EDITOR
Oak Ridge National Laboratory

A. P. Malinauskas, EDITOR
Oak Ridge National Laboratory

G. R. Eidam, EDITOR
Bechtel National, Inc.

H. M. Burton, EDITOR
EG&G Idaho, Inc.

Developed from a symposium sponsored by
the Division of Nuclear Chemistry and Technology
at the 189th Meeting
of the American Chemical Society,
Miami Beach, Florida,
April 28–May 3, 1985

American Chemical Society, Washington, D.C. 1986

Library of Congress Cataloging in Publication Data

The Three Mile Island accident.
(ACS symposium series, ISSN 0097-6156; 293)

"Developed from a symposium sponsored by the
Division of Nuclear Chemistry and Technology at the
189th Meeting of the American Chemical Society,
Miami Beach, Florida, April 28–May 3, 1985."

Includes bibliographies and index.

1. Three Mile Island Nuclear Power Plant (Pa.)—
Congresses. 2. Nuclear power plants—Pennsylvania—
Accidents—Congresses.

I. Toth, L. M. (Louis McKenna), 1941–
II. American Chemical Society. Division of Nuclear
Chemistry and Technology. III. American Chemical
Society. Meeting (189th: 1985: Miami Beach, Fla.)
IV. Series.

TK1345.H37T45   1986   363.1'79   85-26852
ISBN 0 8412 0948 0

ACS Symposium Series

M. Joan Comstock, *Series Editor*

Advisory Board

FOREWORD

The ACS SYMPOSIUM SERIES was founded in 1974 to provide a medium for publishing symposia quickly in book form. The format of the Series parallels that of the continuing ADVANCES IN CHEMISTRY SERIES except that, in order to save time, the papers are not typeset but are reproduced as they are submitted by the authors in camera-ready form. Papers are reviewed under the supervision of the Editors with the assistance of the Series Advisory Board and are selected to maintain the integrity of the symposia; however, verbatim reproductions of previously published papers are not accepted. Both reviews and reports of research are acceptable, because symposia may embrace both types of presentation.

CONTENTS

INDEXES

PREFACE

THE THREE MILE ISLAND ACCIDENT occurred on March 29, 1979. The decision to hold a symposium on the TMI accident aftermath was reached when it was realized that enough information had been gathered during the past 6 years to provide a fairly complete picture of the damage and of the activities required for eventual recovery. The scientific community and the public are generally unaware of these activities; thus a review at this time seems appropriate.

The symposium was organized into three sessions: the first dealt with a description of the accident, the second focused on the chemical aspects involved, and the third addressed the strategy and progress made toward recovery. The symposium was intended to focus on these three subjects and leave the environmental considerations to future meetings and reports. Although it might appear shortsighted to exclude the environmental impact, the exclusion was considered necessary in order to maintain the focus we sought.

This endeavor would not have been successful without the support and cooperation of many individuals and organizations. First of all, Ray G. Wymer (Oak Ridge National Laboratory) and Richard W. Hoff (Lawrence Livermore National Laboratory) should be recognized for the initial impetus on this project. Next, my coorganizers, Tony Malinauskas (Oak Ridge National Laboratory), Greg R. Eidam (Bechtel National, Inc.), and Harold Burton (EG&G) must be identified as the muscle that produced the proper program and the involvement of the best experts on the subjects. Especially significant has been the cooperation of the GPU Nuclear Corporation (which operates TMI) in providing all the necessary staff and technical details. As a result, the papers presented here are from some of the key persons involved in the cleanup operation.

L. M. TOTH
Oak Ridge National Laboratory
Oak Ridge, TN 37831

September 12, 1985

THE ACCIDENT

1

Description of the Accident

Garry R. Thomas

Safety Technology Department, Nuclear Power Division, Electric Power Research Institute, Palo Alto, CA 94303

The TMI-2 accident occurred in March 1979. The accident started with a simple and fairly common steam power plant failure--loss of feedwater to the steam generators. Because of a combination of design, training, regulatory policies, mechanical failures and human error, the accident progressed to the point where it eventually produced the worst known core damage in large nuclear power reactors.

Core temperatures locally reached UO_2 fuel liquefaction (metallic solution with Zr) and even fuel melt (3800-5100°F). Extensive fission product release and Zircaloy cladding oxidation and embrittlement occurred. At least the upper 1/2 of the core fractured and crumbled upon quenching. The lower central portion of the core apparently had a delayed heatup and then portions of it collapsed into the reactor vessel lower head. The lower outer portion of the core may be relatively undamaged. Outside of the core boundary, only those steel components directly above and adjacent to the core (<1 foot) are known to have suffered significant damage (localized oxidation and melting). Other portions of the primary system outside of the reactor vessel apparently had little chance of damage or even notable overheating.

The demonstrated coolability of the severely damaged TMI-2 core, once adequate water injection began, was one of the most substantial and important results of the TMI-2 accident.

Early on the morning of March 28, 1979, at almost exactly 4:00 am, the 880-MWe Three Mile Island-Unit 2 (TMI-2) pressurized water reactor (PWR), which was operating at nearly full power (98% full power or ~2720 MWth), had a loss of feedwater event. What should have been a normal sequence of events responding to the loss of feedwater and resulting in an uneventful shutdown of the plant did not proceed as planned--and the reactor core ultimately was badly damaged. This paper presents both a brief history of the important aspects of the

0097-6156/86/0293-0002$07.00/0
© 1986 American Chemical Society

accident that affected the core damage (1-3) and a summary of the
actual damage as currently known.

Background

A PWR such as the Babcock and Wilcox designed TMI-2 plant has separate primary coolant and secondary coolant systems. The primary
system of a PWR (Figure 1 for the TMI-2 system) is operated at a
sufficiently high pressure--2200 psia (15.2 MPa) for TMI-2--to prevent reaching a bulk saturated water temperature and net steam formation during any normal operation. At full power, the pressurized
water flows upward through the core at a rate of ~1.38 x 10^8 lbm/h
(~1.74 x 10^4 kg/sec) and is heated from 557°F (565 K) to 611°F
(595 K). The water exits the core and reactor vessel and enters the
tube side of the steam generators via large vessel outlet pipes,
which at TMI-2 are called the candy-cane hot legs because of their
characteristic shape (Figure 1). At TMI-2, there are two steam generators with once-through flow on the primary side, as shown for a
single generator in Figure 1. Each steam generator is fed by one hot
leg and is emptied by two cold legs which return water to the reactor
vessel under the driving head of main coolant pumps.
    Steam to drive the turbine-generator unit is formed in the secondary system in the shell side of the steam generators. A complete
TMI-2 system schematic is shown in Figure 2, again with only one of
two loops (the A loop) shown. The A loop contains the pressurizer
and its relief valve that play a major part in the TMI-2 accident.

The Accident

With the loss of feedwater supply to the main feedwater pumps, caused
by a condensate pump trip, the main feedwater pumps also trip. (See
Figure 2 for locations of these and following reactor system components.) The system, still responding normally, trips the turbine
generator and starts the auxiliary feedwater pumps in an effort to
maintain secondary side heat removal in the steam generators. This
is required for maintaining proper primary system temperatures and
pressure until the reactor can be scramed (nuclear power shutdown)
and the reactor system placed in stable standby conditions.
    However, in spite of starting the auxiliary feedwater pumps, no
feedwater supply reaches the steam generators. Block valves in line
with the auxiliary feedwater pumps had been closed earlier as part of
the requirements of a recently-completed test operation of the system--but the valves had inadvertently remained closed after test completion; hence, no feedwater can reach the steam generators.
    We are still in the first few seconds of the accident. Loss of
ability for the primary system to dump its nearly full power load results in an increase in water volume in the primary system because of
heating. This increase is seen directly as a level increase in the
pressurizer located on the A loop. The pressurizer, normally about
half full during full power operation, acts as both the primary
system accumulator plus controller of system pressure. It controls
pressure by maintaining the proper pressure limits in its steam-
filled upper region through a combination of electrical heaters in
its lower water-filled portion and spray cooling in its upper
portion.

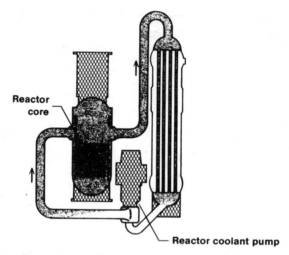

Figure 1. Schematic of the TMI-2 primary system (only one of two nearly identical loops shown). Reproduced with permission from Ref. 1. Copyright 1980, Electric Power Research Institute.

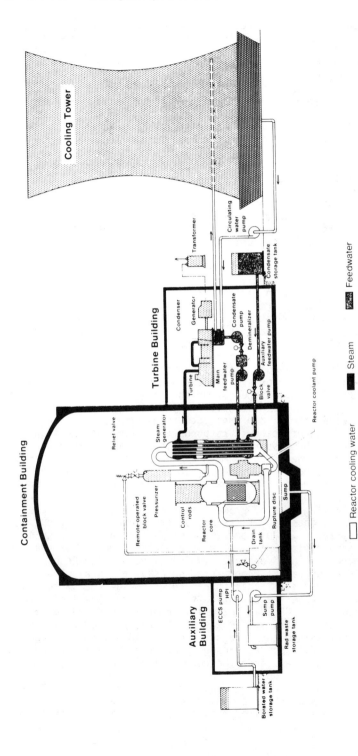

Figure 2. Schematic of the major TMI-2 systems involved in the TMI-2 accident (again, only one loop, the A loop, is shown). Reproduced with permission from Ref. 1. Copyright 1980, Electric Power Research Institute.

   The pressurizer level increase produces a rapid pressure increase which causes both the reactor to scram and a power-operated
relief valve (PORV), mounted at the top of the pressurizer, to open
as an immediate pressure relief (both are normal safety system responses to overpressure).

   It is only nine seconds since the condensate pump trip started
the above chain of events. The entire system has easily weathered
the loss of feedwater event, a fairly common occurrence in steam
power plants whether nuclear or fossil-fired. The reactor has been
shutdown and the excess primary system energy is being adequately
handled even in spite of the total loss of feedwater owing to the
closed valves in the auxiliary feedwater lines. There is no damage
to any major reactor components at this time.

   As a result of the PORV opening to relieve primary system pressure, steam (and possibly some carryover of liquid water) is venting
to the drain tank and system pressure is decreasing. Once the nominal operating pressure is reached, the PORV should close--starting a
period of controlled cooldown until the excess primary system energy
is removed.

   Now the major problems begin that ultimately lead to core damage
(which does not even start until more than two hours later). When
the system pressure falls to the PORV reset valve, the valve fails to
close--for a still unknown reason. This condition is not discovered
because the PORV instrumentation in the reactor control room did
properly monitor the fact that the PORV solenoid energized--an indication that the PORV received and acted upon a closure signal--but
the valve did not close. The control room instrumentation did not
have a valve position readout--a situation that since has been uni-
versally corrected in appropriate nuclear power plants as one of the
very many direct "lessons learned" from the TMI-2 accident.

   With the PORV still open, both primary system pressure and water
inventory are decreasing. When the system pressure reaches ~1650
psia (~11.4 MPa) a safety circuit responds to a low pressure signal
and initiates the high pressure injection system (HPIS)--an engineered safety feature designed to maintain primary system water inventory, and which feeds water directly into the cold legs of the
primary system. The water then flows into the reactor vessel downcomer as added inventory.

   With a continuing loss of primary system coolant and pressure,
steam-filled voids, caused by flashing, start to occur throughout the
TMI-2 primary system. The presence of these voids causes a rise in
pressurizer water level. Such a rise in a Babcock and Wilcox PWR
system was frozen into both operator training and Nuclear Regulatory
Commission (NRC) approved emergency procedures as a signal of excess
water in the primary system and a threat to go solid--the system completely filled with water--a situation to be avoided at all costs.
As a direct result, the HPIS was manually throttled--according to
procedures--and the primary system was on its way to eventually uncovering the core. Since the core still has residual heat generation
of a few percent of full power, caused by decay of radioactive nuclides in the nuclear fuel, once the core begins to uncover, the uncovered portion will heat up.

   The HPIS remained throttled until about 200 minutes (3 h-20 min)
into the accident--by which time, all major core damage that occurred
was completed or on a preordained pathway to completion. This is one

of the big "If only..." conditions of the accident. Despite the
large core size (~37000 fuel rods and ~130 tons) and residual decay
heat of ~31 MW by the time the core began to uncover, only a small
sustained flow of ~200 gpm (~13 kg/sec) of subcooled water by the
HPIS (~25% of its flow capacity) during that 200-minute period would
have prevented any core damage (Table I). As it was, the flow during
this period was generally about 1/3 to 2/3 this value, and at times,
effectively was reduced to zero flow. As a result, the accident at
TMI-2 produced the worst known core damage in the history of commercial nuclear power production.

Table I

Required Coolant Flow Rates for Removing Decay Heat
from 880-MWe Power Plant

Decay Heat (% Full Power)	Approx. Time (Hours)	Minimum Coolant Flow to Core Inlet Plenum[1,2] (Gallons of water per minute)	
		Complete Subcooling	No Subcooling
2%	0.24	340	540
1%	2.6[3]	170	270
0.5%	22	85	135
0.25%	113	43	68

Notes: (1) The core coolant flow rates represent the actual flow
             that is necessary to completely cover the core and re-
             move decay heat by producing steam. The core can be
             about 1/4 to 1/3 uncovered for a sustained period and
             still not be significantly damaged.
        (2) The rates are based on an assumed system pressure of
             680 psia (4.7 MPa) or 500°F (533 K) saturation tempera-
             ture. The pressure is of only minor importance. At
             atmospheric pressure, flow rates would be increased by
             less than 5% for the completely subcooled case and de-
             creased by ~25% for the saturated (no subcooling) case.
        (3) In the midst of rapidly increasing core damage in the
             TMI-2 accident.

We are still in the early phase of the accident. The above actions through the shift to manual control and throttling of the HPIS
occurs before the accident is five minutes old; but the pathway to
both eventual core damage and fission product release from the core
and the primary system has been established. The primary system is
in an unrealized mode of net water inventory loss and the reactor
drain tank (Figure 2), which has been accepting flow from the still-
opened PORV, has filled and its relief valve has opened, feeding the
overflow into the reactor containment building. In a short while,
~15 minutes into the accident, the flow loading to the drain tank
will cause the tank's rupture disc to fail. The combination of the
failed open PORV plus the rupture disc failure provides a direct and
continuously open pathway for primary system inventory loss to a

large receiving volume (the containment building), and eventually, a fission product pathway out of the primary system.

By 100 minutes into the accident, the PORV is still open and so much primary system water had been lost that the two-phase flow through the two main coolant pumps still operating (two had been stopped ~25 minutes earlier) was causing sufficiently severe vibrations in the pumps that they had to be stopped. Upon stopping, phase separation occurs and water fills the lower portions of the primary system, with the reactor vessel filled to above the core top.

It is nearly another 10 minutes before any of the core begins to uncover. There is some makeup water (throttled HPIS water) coming into the system, but as noted earlier, not enough to keep the core covered.

It will still be another ~30 minutes (~140 minutes into the accident) before any major core damage occurs (probable timing of initial fuel rod failures) and it is realized that the PORV is open. To circumvent the open PORV, an in-line block valve (Figure 2) is closed. For the first time, the primary system is closed. But since it still has not been recognized that the open PORV caused partial core uncovering, severe core damage is inevitable. However, it will probably be another 15-20 minutes before severe fuel damage and high core temperatures are reached.

The period of major core damage appears to have lasted from ~140 minutes to ~225 minutes and is discussed in the following sections.

Core Damage

The actual level of core damage at TMI-2 is far more extensive than in any other known power reactor accident. Unfortunately, because of both:

1. limitations in actual recorded data from the accident;
2. characteristics of the damage processes (particularly the highly exothermic Zircaloy oxidation and the processes associated with the recently discovered (4) apparent slumping of part of the lower core into the reactor vessel lower head);

the actual patterns and types of core damage cannot be uniquely established via analyses. Final definition of core damage will have to await detailed examination of the TMI-2 primary system and the core.

Despite data limitations, the existing evidence of core damage does indicate examples of extreme material performance and damage. These examples can be characterized by the following known information related to the damage processes or conditions during the TMI-2 accident.

1. The large quantity of hydrogen gas produced during the accident (~1000 lbm or ~450 kg) is most readily explained by oxidation of a major portion of the core Zircaloy inventory in steam. This information is of limited use in modeling TMI-2 core damage because the time-rate of hydrogen production currently cannot be quantified--i.e., the rate of oxidation-driven damage cannot be defined because of inadequate data on hydrogen production at TMI-2. However, with recent advances in modeling integral primary and secondary system behavior during severe degraded core

accidents (5,6), it is possible to be reasonably confident that only a small portion of the total hydrogen (<20%) was formed prior to a major core quench starting at ~174 minutes (Item 10 below).

2. The exponential relationship between the exothermic Zircaloy oxidation process and Zircaloy cladding temperature (7,8) did drive local core temperatures to very high levels. Because of a low set-point (off-scale) value of 700°F (644 K), the 52 core outlet thermocouples that existed just above the TMI-2 core (at the top of the 52 instrumented out of 177 total fuel assemblies, Figures 3 and 4) could not produce definitive core outlet temperatures for the period that severe damage occurred. The 52 instrumented fuel assemblies were arranged in a spiral pattern from core center to core edge (Figure 4), thereby providing a quite complete spatial coverage of core exit temperatures. With a higher set point, these outlet thermocouples could have provided verification of analytical predictions of core temperatures up to ~2400°F (~1600K).

3. The recorded behaviors of the 52 strings of self-powered neutron detectors (SPNDs--seven per string located in each instrumented fuel assembly, Figure 3) indicate that:
    (a) 2/3 to 4/5 of the axial core height was uncovered (dried out) at times during the period that severe core damage was occurring, as schematically shown in Figure 5 (1,2,3,6,9),
    (b) a delayed event occurred, after the core was recovered with water, that caused sudden failure of several centrally-located SPND instrument strings and apparently is related to a partial core slump into the reactor vessel lower head.

4. The ex-vessel source-range neutron detector or SRND (Figures 3 and 4) provides a relatively definitive but not analytically unique history of the drying out and recovering of the TMI-2 core (Figures 6 and 7 and Table II). Physics analyses of the SRND signal does produce a generally consistent water level history for the first period of core uncovering (~110 to ~174 minutes) compared to integral core heatup analyses (6,9,10). The SPNDs provide additional evidence of this same history, but because of the unpredictable nature of their overheated behavior, interpretation of core dry out history from the SPND data is much less definitive.

5. The timing and magnitude of the response of six independent TMI-2 containment building radiation monitors indicate:
    (a) the approximate timing of initial fuel rod failures caused by the drying out and ensuing heating up of the core -- as is shown in Figure 8, these signals indicate first fuel rod failures ~30 minutes after the core started to uncover, which began ~110 minutes into the accident;
    (b) that the core reached a high state of disruption (high temperature and/or chemical and/or mechanical disruption based on the rapid increase in fission product release from the failed fuel. This release appears to be completed in less than one hour after the first indication that fuel rods had failed (by ~195 minutes), as indicated by the peaking of the radiation monitors (1).

6. Most of the fission products remained in the core. Of those released from the core, most, except for gaseous Xe and Kr,

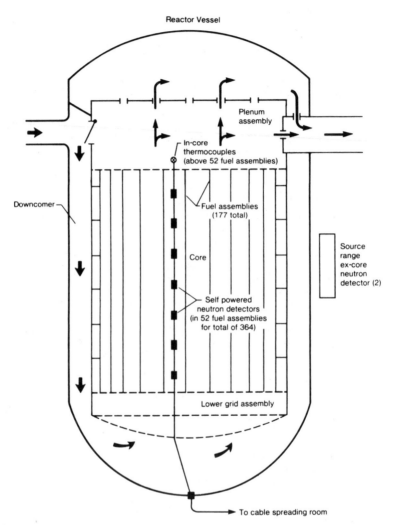

Figure 3. Schematic of the TMI-2 pressure vessel and in-core instrumentation string configurations. Reproduced with permission
from Ref. 1. Copyright 1980, Electric Power Research Institute.

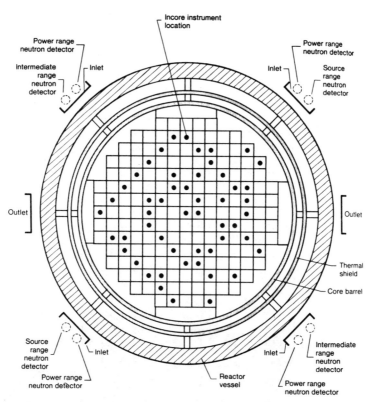

Figure 4. Top sectional view of the TMI-2 reactor pressure vessel showing in-core and ex-vessel instrumentation locations. Reproduced with permission from Ref. 1. Copyright 1980, Electric Power Research Institute.

REACTOR VESSEL

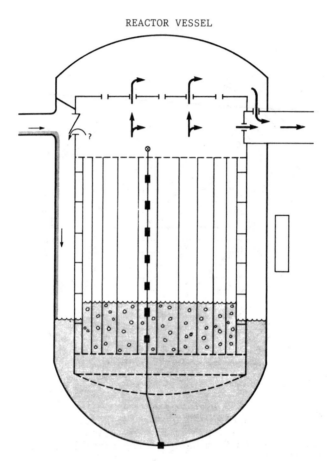

Figure 5. Schematic representation of the minimum water level in the TMI-2 core which probably occurred between 160 and 174 minutes into the accident. Note the limited (throttled) high-pressure injection flow represented on the inlet (left) side of the schematic. Reproduced with permission from Ref. 1. Copyright 1980, Electric Power Research Institute.

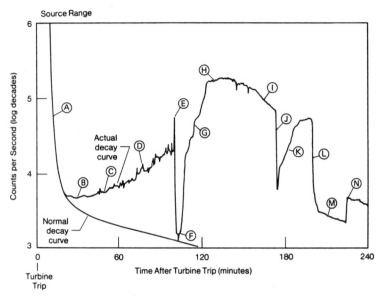

Figure 6.   The ex-vessel source range neutron detector signal during the TMI-2 accident.   See Table II for an explanation of callouts.   Reproduced with permission from Ref. 1.   Copyright 1980, Electric Power Research Institute.

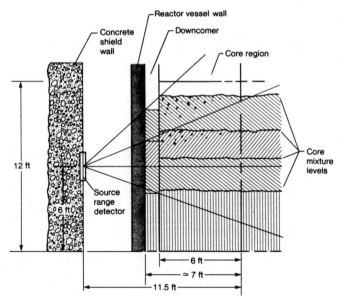

Figure 7.   Source range neutron detector general arrangement and field of view used in the physics modeling for helping to understand the detectors signal (Figure 6).   Reproduced with permission from Ref. 1.   Copyright 1980, Electric Power Research Institute.

Table II

Interpretation of Source-Range Neutron Detector Signal

(Refer to Figure 6 for Locations of Alphabetical Notations)

A. For the first 20 minutes of the accident, behavior was consistent with the response during a normal post-trip decay heat curve.

B. After approximately 20 to 30 minutes, the detector source count rate (signal) began to rise above the expected decay heat curve due to increasing production of voids (steam bubbles) in the downcomer and core regions. The void reduced the shielding effectiveness of the water in these regions. This behavior is consistent with the system reaching bulk saturation ~6 minutes after turbine trip, and net outflow through the open PORV.

C. Continued loss of coolant from the primary system led to increased voidage and an increase detector signal. Signal noise began, indicating unsteady flow (pump surging) and phase separation characteristic of two-phase slug flow. The noise increased with time.

D. At 73-74 minutes, two of the four reactor main coolant pumps were secured by the operators.

E. At 100 minutes, the final main coolant pumps were secured. This caused a separation of voids to the upper regions and liquid water to the lower regions of the system. The detector signal abruptly dropped as a result of the downcomer being filled.

F. The minimum count rate indicates the downcomer was full and the active core covered. (The full downcomer acts as a shutter worth ~100 x in detector signal level.)

G. The core decay heat continued to boil off the water inventory in the core and downcomer, releasing it from the primary system via the still open PORV. Makeup was not sufficient to maintain downcomer and core water levels. As the water levels dropped, the count rate increased.

H. The signal level continued to increase but at a slower rate as shielding variations begin to reach the limit of single-valued signal behavior. Also, the rate of water level drop slowed somewhat in this period because core boil-off tended to balance with the relatively unchanging make-up flow. The open PORV was blocked at ~140 minutes, but condensation in other parts of the primary system still allowed water to leave the reactor vessel.

I. Over this period, the count rate was decreasing. Because of the bivalued nature of the detector for these conditions, the turnaround in detector count rate could be either a core refill or a continued uncovering. However, based on make-up flow estimates (showing a decrease) and other core instrumentation, the decreasing count rate appears to be responding to continued uncovering of the core.

J. Operators started a reactor main coolant pump, sending a slug of cold water into the downcomer and essentially filling it. The downcomer was suddenly filled and the signal drops immediately, but not close to the expected decay heat signal as occurred at point F.

K. Loop flow data indicates that the pump worked effectively for a very brief period (<10 seconds). This is corroborated by the abrupt turnaround in the detector signal. As flow ceased, excess downcomer water moved into and quenched much of the core. This water was then boiled off again from both the quenching plus the continuing decay heat, causing a partial reuncovering of the core.

L. HPIS flow was initiated at 200 minutes. The flow passed into the downcomer, filling it, the addition of HPIS flow began to requench the core. It is conjectured that the coolant first rewetted the outer region of the core, bypassing the hot center.

M. The return to a downward trend with a slope similar to but offset above the normal decay heat curve indicates that the downcomer and core are recovered but the core may be rearranged or still have some voided (hot) regions.

N. The jump in detector signal at ~225 minutes appears to be due to the displacement of fuel downward into the reactor vessel lower head where it is less shielded from the detector.

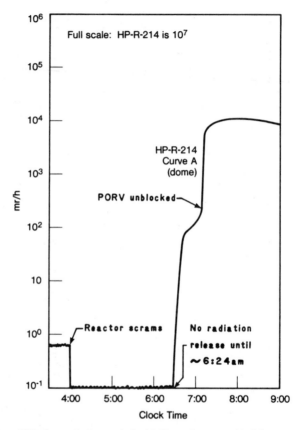

Figure 8. TMI-2 containment building dome radiation monitor signal during the TMI-2 accident. This monitor is mounted well above
the reactor pressure vessel. Reproduced with permission from
Ref. 1. Copyright 1980, Electric Power Research Institute.

ended up in the water on the containment building floor plus in the auxiliary building sump shown in Figure 2 (11).

7. The temperatures in the uncovered and severely damaged regions of the core should have generally exceeded melting temperatures for the control rod materials contained in the core, which consisted of a combination of silver, indium and cadmium. (The TMI-2 core contained ~3 metric tons of these materials.)

8. Core temperatures also generally should have exceeded mechanical failure conditions or melting of stainless steel components in the uncovered portion of the core, which includes the cladding for the control rods.

9. Core temperatures did locally exceed the Zircaloy melting temperature (~3400°F or ~2150 K); and, based on metallurgical examinations of selected particles from the small amount of core debris (~1 lbm or ~0.5 kg) that has been remotely retrieved from the TMI-2 core (12-15), local core temperatures also exceeded the melting points of both the oxide formed from oxidation of a U-Zr-O metallic solution (~4800°F or ~2900 K) and UO_2 (~5100°F or ~3100 K).

10. At ~174 minutes into the accident, one of the main coolant pumps operated for a very short period and virtually threw into the now nearly empty reactor vessel (Figure 5) a large quantity of water (~1000 ft³ or ~30 m³) in less than 10 seconds (1,3,9). This water flow initiated a major quenching of the uncovered and extremely hot upper ~2/3 of the core. The UO_2 fuel is a brittle ceramic and the Zircaloy cladding surrounding the fuel becomes embrittled after being subjected to hot (≥1600 K) oxidizing conditions (7). This sudden infusion of water into the core region probably caused the majority of the visible and extensive damage to the upper core as a result of thermal stress-caused breakup of the upper portion of most of the ~37,000 fuel rods in the core. This damage has been extensively surveyed (12,16-18). From these examinations, it was found that:

    (a) much of the upper core apparently had broken up into small pieces (≤1 cm);
    (b) a large void of ~330 ft³ (~9.3 m³) was left behind in the upper core region as a result of this breakup;
    (c) a large debris bed ~40 inches (~1 m) deep has been found at the bottom of the void with the upper surface of the bed ~4 feet (~1.2 m) below the original top of the active core (the original active core height was 11.8 feet or 3.6 m);
    (d) at the base of the debris bed, there appears to be an extensive layer of a hard material of unknown thickness and composition.

11. At ~200 minutes into the accident, the HPIS had been manually restarted. The original core region was covered and the reactor vessel nearly filled within 10-15 minutes. However, ~10 minutes later, at ~225 minutes (3 h-45 min), there was an apparently large contact between the primary system water and very hot material--based on a wide variety of measurements that showed substantial thermal-hydraulic energy moving throughout the primary system. The amount of energy moving throughout the system was large, but not as large as that accompanying the ~174-minute event described in Item 10 (19).

At the same time, as seen in Figure 6 and Table II, there
was a sudden jump in apparent core nuclear activity as seen by
the SRND. This jump was probably an increase in count rate
caused by UO_2 fuel moving to a less shielded position with respect to the SRND, thereby increasing its effective signal. As
noted above in Item 3, there was a coincidental failure of several centrally-located in-core instrument strings.

This event appears to be slumping of some portion of the
lower central region of the TMI-2 core and some below-core steel
components into the reactor vessel lower head.

12.   Recent ex-vessel radiation measurements and in-vessel remote TV
      video inspections (4,20) show that:
      (a)   a large mass of fuel (probably >10 tons) now resides in the
            reactor vessel lower head;
      (b)   much of this material appears to have a once-molten struc-
            ture;
      (c)   this material has a higher ratio of non-fuel to fuel than
            existed in the original core.

13.   Handmade measurements of the 52 core exit thermocouple (TC)
      signals (using a potentiometer) were taken over a ~1 hour period
      starting a short while after the core slumping into the lower
      head occurred at ~225 minutes (measurements made between four
      and five hours after start of accident). These measurements did
      indicate isolated temperatures near stainless steel melt as
      shown in Figure 9. These measurements have the following
      characteristics:
      (a)   the readings were taken under extreme conditions--in the
            midst of the progressing accident and without a proper
            reference junction;
      (b)   the core exit TC junctions had unknown physical conditions
            and locations after such extensive core damage.
      However, post-accident low-temperature behavior of most of
      the 52 TCs generally appeared to be consistent with expected
      normal behavior, but undetected high-temperature effects may
      have resulted in some form of high-temperature TC decalibration.
      Also, measurements of TC lead lengths made many months after the
      accident showed changes in the resistance in the leads from be-
      fore to after the accident. These changes indicated many of the
      TC junctions had, at some prior time, relocated to the bottom
      head region, particularly most of those with the highest temper-
      atures in the handmade measurements (21).

14.   One of the more unusual aspects of the physical damage caused by
      the uncovering and heatup of the TMI-2 core is the lack of major
      above-core damage. As already discussed, the majority of the
      TMI-2 core is very badly damaged--mainly caused by severe over-
      heating. Yet in the region above the core, the damage is essen-
      tially limited to about 1 to 1.5 feet (~0.3 to ~0.45 m) imme-
      diately above the top of the active core (top of the UO_2 fuel
      columns). Within this region, there is a wide variety of damage
      to small stainless steel or Inconel components (such as the fuel
      assembly grid spacers and upper end fittings). This damage
      ranges from virtually nonexistent to locally gross melting and
      even foaming stainless steel oxidation (occurs very close to
      stainless steel melting conditions in an oxidizing environment).
      That span of local damage, from none to essentially complete,

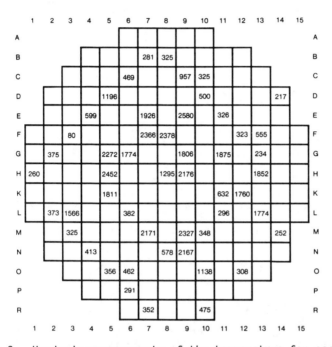

Figure 9. Handmade measurements of the temperature for each core exit thermocouple taken roughly between four and five hours into the TMI-2 accident. Temperatures are shown at the original core position for each thermocouple; however, the exact location and condition of each thermocouple junction was (and is) unknown. Reproduced with permission from Ref. 1. Copyright 1980, Electric Power Research Institute.

(a) appears to occur in a relatively random pattern with the most severe damage concentrated above the mid-radius region of the core, (b) can occur within a few (<5) inches of lateral movement across this region.

Much of the general damage discussed in this section has been anticipated since shortly after the accident (22). The major surprises in the in-vessel remote T.V. video exams (4,16,18) were the lack of damage above the core and the extent and form (apparently once molten) of debris in the reactor vessel lower head.

Summary of Core Damage

Figure 10 (11) presents a summary of the above description of core damage and essentially represents the current understanding of the physical state of the damaged TMI-2 core which is briefly restated below.
1. The large void is shown in the upper core region with:
    (a) a large and loose debris bed at the bottom of the void;
    (b) remnants of fuel assemblies with partial and complete fuel rods radially surrounding the void;
    (c) remnants of above-core components at the top of the void.
2. Just below the debris bed, at about the 40% core height (~56 ±10 inches), is an apparenty hard layer of unknown thickness and composition.
3. In the lower core region, the condition is unknown as there has been no examination of this region.
4. A large amount of fuel and non-fuel material has relocated into the bottom head of the reactor vessel--and much of the visible material does appear to have been once molten.

Long-Term Core Behavior

The singularly most significant aspect of the progressing TMI-2 core damage-reflood-quench sequence was the displayed and recorded coolability of this very severely damaged core, even despite the delayed slumping of apparently molten material into the reactor vessel lower head. Once significant coolant water was injected into the reactor vessel, cooling of the core improved in a nearly monotonic manner.

This improved cooling process is most dramatically displayed by the nearly continuous decrease in fraction of core exit thermocouples that are still above the set-point (off-scale) level of 700°F or 644 K (Figure 11). The cooling started with the short-term operation of a main circulation pump at ~174 minutes which rapidly injected a large quantity of water into the then nearly empty pressure vessel downcomer. The later restart and sustained operation of the HPIS, at about 200 minutes, provided a continuing flow of water to cover the core and keep it covered. From 174 minutes onward, core coolability was improving.

For the period of ~5 to 15[+] hours, there was limited improvement in coolability as measured by the fraction of off-scale TCs in Figure 11. During this period, the primary system was being purposely and extensively depressurized in an attempt to start the low-pressure injection system (another engineered safety feature). This involved the period from ~7.6 hours to ~14 hours. Repressurization and a restart of a main coolant pump between 15 and 16 hours reduced by a

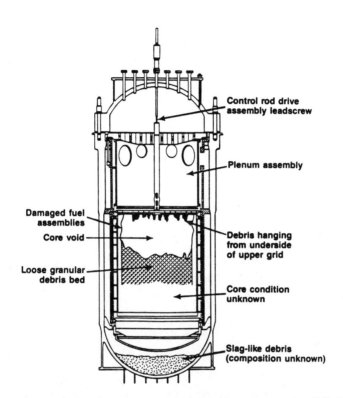

Figure 10. Summary of currently known core damage at TMI-2. Note that the bottom penetrating instrumentation strings and their supporting components are only schematically represented at the outer surface of the bottom head and only one of 67 control rod drive assemblies is shown. Reproduced with permission from Ref. 11. Copyright 1985, D.E. Owen.

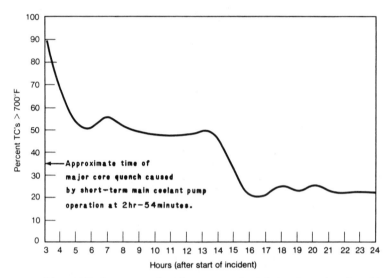

Figure 11. TMI-2 core exit thermocouple histories showing the remaining fraction of thermocouples still off scale (>700°F) as a function of accident time. Reproduced with permission from Ref. 1. Copyright 1980, Electric Power Research Institute.

factor of ~2 the number of TCs still off-scale. Conditions were then well in hand, and shortly, peak system temperatures were very close to local water saturation temperatures. Figure 12 shows this by comparing the hand measured temperatures previously shown in Figure 9 (now in °C) with a set of temperatures taken about one week later.

The accident itself was over, but the controversy, the lessons learned, and the new insight into severe core damage behavior were just beginning.

TEMPERATURE PROFILE IN TMI CORE (IN °C)
MARCH 28, 1979
(APRIL 6, 1979)

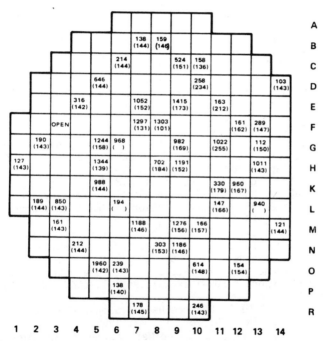

Figure 12.   TMI-2 core exit thermocouple temperatures during the accident (repeat of Figure 9 handmade measurements but now in °C) and nine days later. Reproduced with permission from Ref. 10. Copyright 1982, Electric Power Research Institute.

Literature Cited

1. "Analysis of Three Mile Island-Unit 2 Accident, NSAC-80-1
    (NSAC-1 Revised)", Nuclear Safety Analysis Center (NSAC),
    Electric Power Research Institute (EPRI), Palo Alto, CA, March
    1980.
2. "The Report of the President's Commission on the Accident at
    Three Miles Island--The Need for Change: The Legacy of TMI"
    (Kemeny Commission Report), Washington, DC, October 1979.
3. "Three Mile Island--A Report to the Commissioners and to the
    Public" (Rogovin Report), Vol. II, Part 2, pp. 472-550,
    Washington, DC, February 1980.
4. Fricke, V.R., et al., "Reactor Lower Head Video Inspection,"
    TMI-2 Technical Planning Bulletin (TPB) 85-6, Rev. 1, GPU
    Nuclear Corp.-Bechtel National, Inc., March 1985.
5. "Technical Report 16.2-3 MAAP Modular Accident Analysis Program
    User's Manual"--Volumes I and II, Fauske and Associates, Inc.
    for the Industry Degraded Core Rulemaking (IDCOR) Program, Aug-
    ust 1983.
6. Kenton, M.A., et al., "Simulation of the TMI-2 Accident Using
    the Modular Accident Analysis Program Version 2.0," Fauske and
    Associates, Inc. Report FAI/85-25 (also to be published as an
    EPRI report), Burr Ridge, IL, April 1985.
7. Chung, H.M. and Kassner, T.F., "Deformation Characteristics of
    Zircaloy Cladding in Vacuum and Steam under Transient-Heating
    Conditions: Summary Report," NUREG/CR-0344 (ANL-77-31), Argonne
    National Laboratory (ANL), Argonne, IL, July 1979.
8. Biederman, R.D., et al., "A Study of Zircaloy 4--Steam Oxidation
    Kenetics," EPRI Reports NP-225 and NP-734, Part 2, EPRI, Palo
    Alto, CA, September 1979 and April 1978.
9. Ardron, K.H. and Cain, D.G., "TMI-2 Accident Core Heat-Up Analy-
    sis," NSAC-24, NSAC, EPRI, Palo Alto, CA, January 1981.
10. Warren, H., et al., "Interpretation of TMI-2 Instrument Data,"
    NSAC-28, NSAC, EPRI, Palo Alto, CA, May 1982.
11. Owen, D.E., et al., "Fission Product Transport at Three Mile
    Island," Paper for presentation at the American Ceramic Society
    1985 Annual Meeting, May 1985.
12. Akers, D.W. and Nitschke, R.L., "TMI-2 Core Debris Grab Sample
    Quick Look Report," EGG-TMI-6531, Rev. 1, EG&G Idaho, Inc.,
    March 1984.
13. Hayner, G.O., "TMI-2 H8A Core Debris Sample Examination Final
    Report," GEND*-INF-060, Vol. II, May 1985.
14. Hayner, G.O., et al., "Detailed Metallographic and Microchemical
    Examination of the H8A Core Debris Sample from TMI-2," EPRI Re-
    port to be published, Babcock and Wilcox, Summary 1985.
15. "Summary of TAAG Meeting on TMI-2 Fuel Conditions (March 25-26,
    1985)," GPU Memorandum TMI-2 4000-85-S-151, March 1985.
16. Worku, G., "Plenum Underside Damage," TPB 85-15, Rev. 0, GPU-
    Bechtel, June 1985.

*Consortium of GPU Nuclear Corporation, EPRI, Nuclear Regulatory
Commission and Department of Energy.

17. Beller, L.S. and Brown, H.L., "Design and Operation of the Core Topography Data Acquisition Sysstem for TMI-2," GEND-INF-012, May 1984.
18. Fricke, V.R., "Core Debris Bed Probing," TMI-2 TPB 84-8, Rev. 1, GPU-Bechtel, February 1985.
19. Henry, R.E. of Fauske and Associates, Inc., Personal Communication.
20. "Data Report--Determination of Fuel Distribution in TMI-2 Based on Axial Neutron Flux Profile," TPO/TMI-165, Pennsylvania State University for GPU-Bechtel, May 1985.
21. Yancey, M.E. and Wilde, N., "Preliminary Report of TMI-2 In-Core Instrument Damage," GEND-INF-031, January 1983.
22. Thomas, G.R., Proceedings for the International Conference on World Nuclear Energy--Accomplishments and Perspectives (American Nuclear Society, Volume 35, 1980, pp. 194-6.

RECEIVED July 29, 1985

2

Thermal Hydraulic Features of the Accident

B. Tolman, C. Allison, S. Behling, S. Polkinghorne, D. Taylor, and J. Broughton

Idaho National Engineering Laboratory, EG&G Idaho, Inc., Idaho Falls, ID 83415

The TMI-2 accident resulted in extensive core damage
as confirmed by recent video data showing extensive
prior molten core debris in the bottom of the reactor
vessel. A hypothesized TMI accident scenario is presented that consistently explains the TMI data and is
also consistent with research findings from independent severe fuel damage experiments. The TMI data
will prove useful in confirming our understanding of
severe core damage accidents under realistic reactor
systems conditions. This understanding will aid in
addressing safety and regulatory issues related to
severe core damage accidents in light water reactors.

TMI-2 characterization data over the past 2 years confirm that extensive damage to the reactor core resulted from the TMI accident. To
understand the mechanisms resulting in such extensive core damage, it
is necessary to reconstruct the important thermal hydraulic events
which controlled the accident progression. The previous paper
(1) summarizes the sequence of plant events that led to eventual loss
of reactor vessel coolant, core heatup and eventual failure, and the
subsequent efforts to re-establish long term forced cooling to the
damaged core. This paper discusses the thermal hydraulic features
of the accident that directly influenced the core damage progression
during the period from 100 to 300 minutes (all times unless noted
are relative to the beginning of the accident). A hypothesized accident progression scenario in which the core first attained a non-
coolable geometry and subsequently progressed to a final coolable
state is discussed. Our understanding of the accident progression
scenario and associated thermal hydraulics will be presented based
on applicable TMI data, the results of independent severe fuel
damage experiments, and first order thermal hydraulic principles.

Summary of the End-State Core and Reactor Vessel Conditions

Prior to discussing the thermal hydraulic aspects of the accident,
it is necessary to summarize the damage to the reactor vessel

0097-6156/86/0293-0026$06.00/0

environs and the known end-state conditions of the core materials.
The thermal hydraulics of the core damage progression must then be
consistent with this data. The known state of the reactor vessel
is summarized in Figure 1. Approximately one-third of the original
fuel in the upper core region is no longer there, leaving a void
region extending to the outermost fuel assemblies. Measurements
from the control-rod drive structures indicate that upper plenum
structure temperatures (above the core) ranged from 700 to 1350 K,
several hundred degrees below the stainless steel melting temperatures (2). A debris bed ranging from 2 to 3 feet deep is resting
on top of the existing core. Characterization of particles from
the debris indicate that the zircaloy is highly oxidized, and localized temperatures exceeding 2900 K were reached, sufficient to melt
the UO_2 fuel (3). Efforts to probe down through the remaining
core materials indicate that an impenetrable layer of material
exists at the 5- to 6-foot elevation (above the bottom of the original core). Recent video scans of the lower regions of the reactor vessel indicate that 5 to 20% of the core fuel and/or structural
materials now reside on the vessel bottom, nearly 7 to 8 feet below
the bottom of the original core. The extent of damage to the lower
core support region is not presently known since the video scans
were unable to characterize the center regions of the lower plenum.
Based on the video information, which is limited to the peripheral
regions of the reactor vessel, the reactor vessel structure does
not appear to be significantly damaged, even though as much as
20 tons of core material may reside on the bottom of the vessel.

Overview of the Core Degradation Period (100 to 300 Minutes)

Establishing an acceptable accident scenario based only on TMI data
would be difficult since the available characterization data relate
primarily to the end-state condition of the core and reactor system.
Limited plant data were taken during the accident that relate
directly to the core damage progression. Independent severe fuel
damage experiments, however, provide an important key for interpreting the end-state TMI data. Also, severe core damage computer
models are helpful to interpret and integrate both the end-state
characterization data and the available transient response data
recorded during the accident. Using these resources, together with
the TMI data, a more complete understanding of the core damage progression is emerging.

The time duration over which the major core degradation occurred can be subdivided into three intervals:

1. Initial core heatup and fuel/cladding melting/liquefaction
    from 100 to 174 minutes.

2. Attainment of a non-coolable core geometry and subsequent core
    heatup to above 2400 K from 174 to 227 minutes.

3. Melt progression into the lower regions of the reactor vessel
    and subsequent attainment of a coolable geometry from 227 to
    300 minutes.

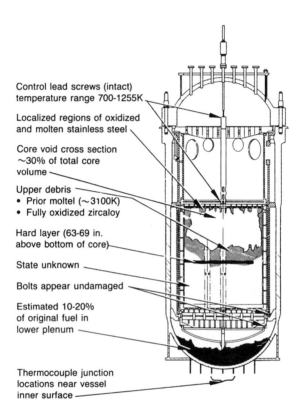

Control lead screws (intact)
temperature range 700-1255K

Localized regions of oxidized
and molten stainless steel

Core void cross section
~30% of total core
volume

Upper debris
• Prior moltel (~3100K)
• Fully oxidized zircaloy

Hard layer (63-69 in.
above bottom of core)

State unknown

Bolts appear undamaged

Estimated 10-20%
of original fuel in
lower plenum

Thermocouple junction
locations near vessel
inner surface

Figure 1.   Known state of the core and reactor vessel components.

The significant thermal hydraulic features of these three time
periods are discussed below.

Initial Core Heatup and Fuel/Cladding Melting/Liquefaction (100 to 174 Minutes)

When the last primary coolant pump was shut off at 100 minutes,
forced coolant convection to the reactor core was lost, and the
liquid remaining in the reactor vessel and vessel outlet piping
(hot legs) settled rapidly into the reactor vessel. The resulting
reactor vessel liquid level is not known exactly, although it has
been estimated to be near or slightly above the top of the core
region (4). In addition, considerable uncertainty exists in the
reactor system makeup flows during this period. The coolant in the
reactor vessel subsequently boiled off during this period as a
result of the core decay powers levels (100 to 200 MW). The initial reactor vessel liquid level and the system makeup flows under
these conditions control the ability to cool the core as will be
seen shortly.

During this period, measurements of superheated temperatures in
the reactor vessel outlet piping (hot legs) and radiation release
from the primary cooling system give critical timing information on
the reactor vessel boildown and core heatup rates. Additional
information relating to the core temperature response and the reactor vessel liquid levels may also be inferred from the incore thermocouples, incore neutron detectors, and the excore source range monitors. Of particular importance are the data trends from the incore
instrumentation. These instruments included 364 self-powered neutron
detectors (SPNDs) and 52 thermocouples. These instruments were
placed in instrument assemblies, each having seven SPNDs and one
thermocouple and were located at different axial elevations. The
instrument assemblies were inserted in the core region from the
bottom of the reactor vessel. The configuration of the instrument
assemblies within the reactor core are shown in Figure 2. Extensive
analysis and interpreting of these instruments are documented in
Reference 5.

A general overview of the core thermal response during this time
interval based on the available instrumentation data is summarized
below:

1. Initial hot leg superheat was detected at 113 minutes (4-7).
   The steam superheat implies that initial core uncovery and
   heatup occurred prior to 113 minutes.

2. Large increases in containment radiation levels occurred at
   142 minutes. The increased radiation resulted from the initial
   fuel rod failures (cladding burst). The core-to-containment
   transport time has been estimated to be from one to several
   minutes (5). Thus, initial fuel failure occurred several
   minutes prior to 142 minutes.

3. At 135 minutes, the upper core temperatures are estimated to be
   in the range of 810 to 860 K based on data from the incore
   neutron detectors (6).

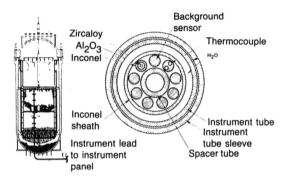

In-core Instrumentation Configuration

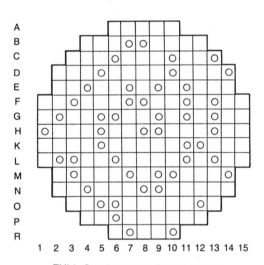

TMI In-Core Instrument String Locations

Figure 2. Incore instrumentation configuration and spatial distribution of instrument assemblies within the core.

4. At 150 minutes, the core temperatures in the upper one-third of the core are inferred to be in the range of 1700 to 1800 K based on data from the incore neutron detectors (6).

5. During the interval of 150 to 174 minutes, the ex-vessel source range monitor (neutron detector) suggests severe damage to the fuel (6). In addition, severe damage is suggested by the incore neutron detectors located in the upper two-thirds of the core region.

To aid in interpreting this set of observations, the Severe Core Damage Accident Package (SCDAP) computer code has been used to estimate the core heatup and degradation (8). The code was developed to model the physical processes controlling severe core damage behavior and has been verified against the available fuel degradation experimental data base (9). Table I summarizes the model features of the SCDAP code.

To estimate bounding limits on the core heatup, calculations were performed for three assumed reactor vessel coolant level histories. The bounding cases are:

1. Nominal Core Uncovery Case. The reactor coolant level vs time was taken from Reference 4. The nominal boundary conditions indicate the reactor vessel coolant level to be near the top of the core at approximately 110 minutes, 10 minutes after the primary coolant pumps were shut off.

2. Early Core Uncovery Case. These boundary conditions assumed the same reactor vessel boildown rates as the nominal case, except the reactor vessel liquid level vs time is shifted 13 minutes to result in an earlier core heatup. The early core heatup results in predicted core temperatures more consistent with measured superheat in the reactor vessel outlet piping by 113 minutes.

3. Early Core Uncovery With Rapid Boildown. This case is the same as the early uncovery case but with no makeup coolant injection into the primary cooling system during the accident. This case represents the most rapid core boildown rate thought to be possible.

Figure 3 compares the estimated core liquid level vs time for each of these three cases. The predicted peak core temperatures are compared to the TMI data in Figure 4.

Three important trends from the SCDAP calculations are noted from the results shown in Figure 4.

1. The core temperature history is very dependent on the assumed reactor vessel hydraulic conditions.

2. The peak fuel rod temperature increases rapidly once cladding temperatures of 1500 to 1700 K are reached. This is a result of the exothermic chemical reaction between the steam and zircaloy fuel rod cladding at these temperatures.

Table I.   Summary of SCDAP Physical Models

General code capabilities.

    Hydrodynamics

      TRAC-BD1/CHAN, quasi-equilibrium (1-D only)

    Radiation

      TRAC-BD1/CHAN, generalized

    Loss of bundle geometry

      Liquefaction
      Ballooning
      Fragmentation

    Fuel rod behavior

      Oxidation (Pawel, Urbanic kinetics, Chung H_2)
      Nuclear heat generation
      Deformation (axisymmetric and non-axisymmetric)
      Fission product release-PARAGRASS (fuel)
      Gap release (Lorenz Model)
      Liquefaction and relocation--ZrO_2 melting, UO_2
      dissolution, ZrO_2 breach
      Fragmentation--Kassner and Chung embrittlement

    Control rod--Zr/SS/Ag-In-Cd

      Oxidation--Zr/SS
      Nuclear heat generation
      Heat conduction
      Liquefaction and relocation

    Shroud--Zr/SS/Other

      Oxidation--Zr
      Nuclear heat generation
      Heat conduction
      Liquefaction and relocation--Zr

    Debris

      Oxidation
      Nuclear heat generation
      Heat conduction
      Fission product release (PARAGRASS)
      Liquefaction

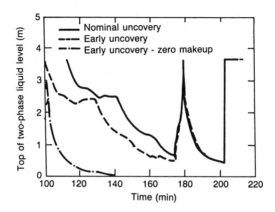

Figure 3. Estimated core liquid levels for three cases of core uncovery.

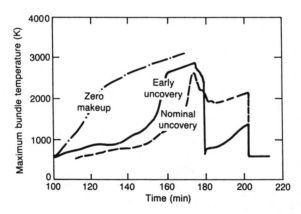

Figure 4. SCDAP predicted peak core temperatures for the three core uncovery cases.

3. For all three cases, the zircaloy melting temperatures are
    exceeded. Thus, it is expected that significant cladding
    relocation (downward flow) would occur during this period.
    This downward flow of molten cladding would be expected to
    interact with the coolant in the lower regions of the core
    causing the zircaloy to freeze thus blocking the coolant flow
    paths through the core.

In addition to the clad melting, the molten zircaloy tends to
react with the UO_2, through oxidation, primarily along the grain
boundaries. This interaction dissolves the UO_2 in the liquid zircaloy and allows grains of UO_2 to be carried away by the molten
zircaloy. This process is referred to as fuel liquefaction. The
extent of fuel liquefaction is controlled primarily by the wetting
behavior of the UO_2/Zr materials which in turn is affected by rod
temperature, duration of the fuel/zircaloy contact, and the geometry of the deformed cladding tubes. Reference 10 discusses the
liquefaction process and data describing this process in more
detail.

Separate effect experiments provide data for understanding the
importance of the zircaloy melting and fuel liquefaction processes.
Figure 5 shows the end-state configuration of a nine rod cluster of
UO_2 fuel rod segments from an experiment with a heatup transient
similar to the SCDAP predicted early uncovery case (11). Extensive
relocation of the zircaloy is evident which results in nearly complete blockage of the coolant flow channels at the bottom of the
fuel assembly. Also, recent experiments conducted in the Power
Burst Facility under simulated TMI core heatup conditions show
extensive cladding melting and fuel liquefaction (12). Extensive
flow blockage in the lower regions of the PBF fuel assembly
resulted from the relocated molten material.

The SCDAP code models the fuel liquefaction, cladding melting,
and relocation processes. The predicted zones of highly oxidized
but intact fuel rods and the zones in which molten and/or liquefied
fuel are predicted are compared for the "nominal" and "early"
uncovery cases in Figure 6. The upper one-third of the core for
both cases is highly oxidized. Molten zircaloy and liquefied fuel
are predicted in the mid-core regions and would have relocated
(flowed) downward, freezing near the coolant interface between the
1- to 2-foot level in the core. It is clear from Figure 6 that the
extent of liquefied material in the lower core regions is dependent
on the reactor vessel hydraulic scenario (boildown rate).

The data trends of the incore neutron detectors during the 150
to 174 minutes period confirm that the melting/liquefaction and
downward relocation of the molten material did occur. The centrally
located incore neutron detectors in the upper third of the core
showed very high anomalous output currents at about 150 minutes.
Best-estimate engineering interpretation of the neutron detector
data suggests the probable explanation of the anomalous, large
positive currents is melting of the instrument sheath. By
165 minutes, neutron detectors in the mid-core region also showed
the anomalous output, suggesting severe core damage in these
regions. If this interpretation is correct, by 170 minutes core
temperatures above 1700 K were reached in the central regions of the
core as low as 30 inches from the core bottom (Level 2 instruments).

Figure 5. End-state condition of fuel rods from recent Germal fuel rod meltdown experiment, ESBU-1. Reproduced from Ref. 5.

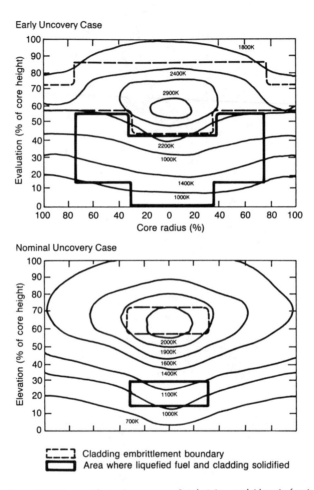

Figure 6. SCDAP predicted zones of highly oxidized (embrittled)
and liquefied cladding for TMI.

The measured reactor system pressure increase from 800 to
1000 psi during the period of 140 to 174 minutes (see Figure 7).
This increasing pressure level indicates severe core damage with an
accompanying large hydrogen production from the zircaloy oxidation
process and/or molten core materials relocating downward and inter-
acting with the coolant in the bottom of the reactor core. Without
severe core damage, the system pressure would not be expected to
increase but rather would decrease as the reactor vessel coolant
level recedes into the lower core regions, where the decay heat is
much less.

The available TMI observations, the best-estimate core damage
predictions, and separate effects core degradation experiments all
indicate that the core experienced significant melting/liquefaction
and downward relocation of the molten core materials prior to
174 minutes. As a result, a molten zone of zircaloy and liquefied
fuel is estimated to exist across nearly 60 to 80% of the core
radius to a height of 0.5 to 1.0 meters. The bottom of the
liquefied zone was supported by a crust of frozen, prior-molten
core material formed at the liquid interface in the bottom
one-fourth of the reactor core. The upper regions of the core
still maintained the original rod-like geometry; however, the
center and part of the upper one-third of the cladding had melted
and relocated downward leaving columns of fuel pellets. The
condition of the fuel rods within the upper and mid-core regions
are expected to be similar to those shown in Figure 5. Any intact
zircaloy cladding in the upper regions of the core would be highly
oxidized and very brittle. Figure 8 summarizes the hypothesized
core conditions just prior to the 'B pump' transient which occurred
at 174 minutes.

'B' Pump Transient and Subsequent Core Heatup (174 to 227 Minutes)

The progression of the accident was significantly altered at
174 minutes, when one of the 'B' loop primary cooling pumps was
started and remained running for approximately 15 minutes. The
coolant delivery to the reactor vessel and core region as a result
of the pump transient is not known. Original estimates (4)
indicate as much as 1000 cubic feet may have been pumped into the
reactor vessel. However, this estimate assumed an intact core
geometry and no flow blockage. For highly degraded core conditions
with significant core flow blockage, coolant flow through the core
would have been limited and would not have been sufficient to arrest
the heatup of the central, molten core region. Figure 7 shows the
reactor system pressure to increase by several hundred psi in the
first minute or so after the pump was turned on. This limited pressurization suggests restricted core flow and heat transfer which is
consistent with extensive core flow blockage and the inability to
cool the hot core with the limited coolant flow. The coolant flow
to the steam generator would also have decreased the system pressures. The relative importance of these interactions is not
presently known and can only be estimated by reactor systems
calculations that model the complex thermal hydraulic interactions
between the degraded core and the reactor system hydraulics.

The mechanical forces and fuel rod thermal stresses generated as a result of the rapid steam generation and rod cooling would

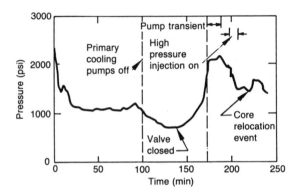

Figure 7.   TMI system pressure history.

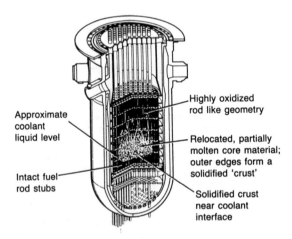

Figure 8.   Hypothesized TMI core configuration just prior to the 'B' pump transient at 174 minutes.

have shattered the oxidized, embrittled fuel rods in the upper core
region. This conclusion is substantiated by data from a recent
32-rod, severe-core damage experiment conducted in the Power Burst
Facility. The data indicate that nearly complete shattering and
relocation of the upper embrittled portions of the fuel rods
resulted from rapid reflooding of the experiment (14).

The hypothesized core configuration shortly after the 'B' pump
transient is shown conceptually in Figure 9. The molten/liquefied
zone would still be in nearly the same location as before the pump
was initiated, since flow upward through the molten material would
be limited because of extensive core flow blockage prior to 174 minutes. The shattered fuel from the upper core region formed a debris
bed approximately a meter deep on or near the top of the molten/
liquefied zone. This upper debris bed would tend to insulate the
center region.

Heat conduction calculations have been performed to investigate
the coolability of a molten/liquefied zone of core material as hypothesized in Figure 9. A simple slab geometry model was utilized and
initial sensitivity calculations varying the assumed slab composition, thickness of the liquefied/molten zone, and surface heat
transfer confirm that the estimated TMI melt zone (center region to
80% radius and from 20- to 30-inches high) would not be coolable,
i.e. the surface heat transfer would not be sufficient to maintain
the center region below the melting temperatures. For this non-
coolable configuration, the molten zone would continue to heat and
eventually melt through the bottom crust and flow into the lower
plenum regions of the reactor vessel.

The TMI incore neutron detectors showed no further anomalous
behavior during the 174 to 224 minute period suggesting that no further melt progression occurred during this period. The incore
thermocouples show a general cooling trend immediately after the
pump transient followed by a slow heatup trend starting at about
190 minutes.

During this period, intermittent high pressure emergency core
cooling injection was initiated, and there was a brief period when
the block valve was opened. As a result of these actions, estimates of the reactor vessel liquid level vs time and the core temperature response is not known and can only be estimated with
mechanistic computer code calculations as indicated earlier.

Melt Progression Into the Lower Regions of the Reactor Vessel and Subsequent Attainment of a Coolable Geometry (227 to 300 Minutes)

At 227 minutes, 53 minutes after the 'B pump transient', a global
change in the core condition occurred as indicated by the incore
instrumentation, reactor system pressure and temperatures, source
range detector, and ex-vessel radiation monitors. Of particular
interest are the incore neutron detectors in the central lower core
regions which indicated anomalous behavior (inferred core melting)
for the first time during the accident at 227 minutes. For many of
the instrument strings (particularly near the core center) this
anomalous behavior was recorded at all axial elevations. In
addition, the lowest incore neutron detectors indicated for the
first time during the accident large changes in output. A summary

of the incore neutron detectors which indicated a significant change
in output is shown in Figure 10.

These data suggest a significant change in core configuration
and damage state occurred at 227 minutes. It is hypothesized that
at this time the melt zone, as shown in Figures 8 and 9, is heated
sufficiently to melt and/or fail the bottom crust or core support
structure. The molten core material then either slumped or flowed
into the lower core and lower plenum regions. The details of the
melt progression and coolability of the molten core material after
failure of the lower support structure is not known.

The flow of molten core material into the lower plenum region
may have been impeded somewhat by the lower core support structures
and elliptical flow distributor plate in the lower plenum region.
However, it can also be hypothesized that the large superheat of the
molten material would have allowed the molten material to flow relatively unimpeded into the lower plenum regions, settling rapidly
onto the reactor vessel. Scoping calculations indicate that if the
core material were to rapidly flow downward onto the reactor vessel,
melt-through of the vessel wall would occur within several minutes
(15). Thus, the intact TMI reactor vessel suggests that (1) the
progression of the core melt occurred over a longer period of time;
(2) the fuel/coolant interaction is important in limiting the downward relocation of the molten core material; and (3) the outer surfaces of the reactor vessel were cooled sufficiently to prevent
failure of the vessel.

The lower plenum video information indicates that instantaneous
relocation of the molten material onto the reactor vessel did not
occur as suggested by:

1. Frozen, prior-molten material on the upper side of the flow
    distributor plate, indicating that the lower plenum structures
    were effective in slowing the melt progression.

2. A variety of frozen material structures (sizes, shapes,
    material, and texture) suggest that different materials pene-
    trated the plenum regions possibly at different times.

The trends from the incore thermocouples give additional
insight into the coolability of the molten core material as it penetrated the lower plenum regions. The thermocouples were melted during the course of the accident and formed new junctions in the
lower, cooler regions of the reactor vessel. Thermocouple measurements from the relocated junctions made between 240 to 330 minutes
indicate temperatures above 800 K (>1000 F) in the center regions
of the reactor vessel as shown in Figure 11. Resistance measurements made on the thermocouple leads after the accident indicate
that for most of the centrally located thermocouples, the relocated
junctions are located near the bottom of the reactor vessel. These
measurements are consistent with the known severe core damage and
indicate the lower plenum fuel debris was above 800 K for times
greater than approximately 2 hours after the "227 minute" core
relocation event.

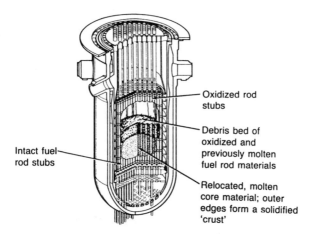

Figure 9. Hypothesized TMI core configuration just after 'B' pump transient (175 to 180 minutes).

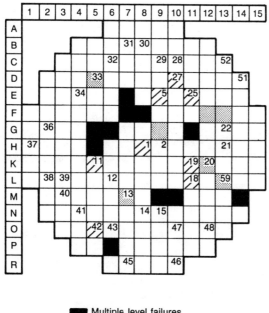

Figure 10. Summary of incore neutron detector locations which indicated a change in core conditions at 227 minutes.

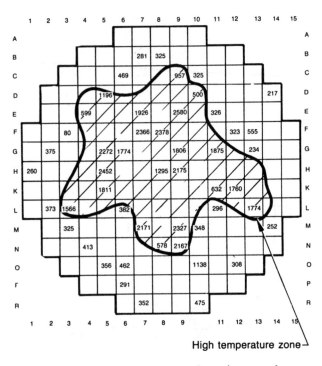

Figure 11. Temperature measurements from incore thermocouples recorded at 240 to 330 minutes.

Conclusions and Summary

A hypothesized TMI accident progression scenario has been described
that interprets the TMI data and is consistent with our understanding of severe core damage phenomena gained from separate
effects experiments. Although many questions remain, a more complete understanding of the accident is emerging. The proposed
accident scenario serves as a basis for interpreting the results
and conclusions presented in papers to follow in the Symposium.
     The major thermal hydraulic features that controlled the
accident are:

1.   Severe core melting and/or fuel liquefaction occurred prior to
     the 'B' pump transient. The initial molten material (zircaloy
     cladding and some fuel) formed a crust in the lower regions of
     the core of sufficient thickness to support the core materials
     as the core melt progressed. The molten core regions could
     have extended over 60 to 80% of the core radius to a height of
     20 to 30 cm. This material would not have been coolable based
     on our initial scoping studies.

2.   The 'B pump transient' caused extensive shattering of the
     brittle fuel in the upper core region and resulted in a debris
     bed 50 to 70 cm high which further tended to insulate the
     center molten region. As a result, the molten/ liquefied zone
     continued to heatup and eventually melted through the lower
     supporting crust at 227 minutes.

3.   Propagation of the molten core material continued into the
     lower plenum after 227 minutes. The interaction of the molten
     material with the coolant in the lower plenum regions is not
     certain; however, the melt progression appears to have been
     impeded by the lower plenum structures. Final cooling times
     may have required several hours.

The TMI data form an important baseline for confirming our understanding of severe core damage phenomena, particularly in light of
molten material challenge to the reactor vessel. The data provide
the necessary benchmark for assessing the capability of the analytical models to predict the controlling core degradation phenomena
during a severe accident in a large reactor system environment.
The TMI data will also provide the data for evaluating the typical-
ity of smaller scale, severe accident experiments. Additional work
is necessary to characterize the lower plenum, the core regions,
and the reactor system flow paths to provide the needed data for
understanding the melt progression into the lower regions of the
reactor vessel and the fission product behavior during the accident.
This data will provide a realistic basis for judging our understanding of severe accidents and their relationship to reactor
safety.

Acknowledgments

This work was supported by the U.S. Department of Energy Assistant
Secretary for Nuclear Energy, Office of Remedial Action and Terminal
Waste Disposal, under DOE Contract No. DE-AC07-76ID01570.

Literature Cited

1.   Thomas, G., "Description of the TMI-2 Accident," These
     proceedings, May 1985.
2.   Vinjamuri, K. et al., Examination of H8 and B8 Leadscrews from
     Three Mile Island, Unit 2, EG&G Idaho, Inc.
     Report EGG-TMI-6685, October 1984.
3.   Akers, D., et al., Draft Report: TMI-2 Core Debris Grab
     Samples--Examination and Analysis, EGG-TMI-6853 (Part 1),
     July 1985.
4.   Ardron, K., Cain, D., "TMI Accident Core Heatup Analysis,"
     NSAC-24, January 1981.
5.   NSAC-28, "Interpretation of TMI-2 Instrument Data," May 1982.
6.   Taylor, D., EG&G Idaho, Inc., personal communication.
7.   NSAC-80-1, "Analysis of Three Mile Island Accident,"
     March 1980.
8.   Allison, C. et al., "SCDAP/MOD1 Analysis of the Progression of
     Core Damage During the TMI-2 Accident," EG&G Idaho, Inc.
     Report SE-CMD-84-006, July 1984.
9.   Allison, C. et al., "Draft Preliminary Report for Comment:
     SCDAP/MOD1/Theory and Models," EG&G Idaho, Inc. Report
     FIN A6360, January 1985.
10. Hofmann, P., Kerwin-Peck, D., and Nikolopoulos, P., "Physical
     and Chemical Phenomena Associated with the Dissolution of
     Solid UO_2 by Molten Zircaloy-4," Zirconium in the Nuclear
     Industry: Sixth International Symposium, ASTM STP 824,
     D. G. Franklin and R. B. Adamson, (eds.), American Society for
     Testing and Materials, 1984, pp. 810-834.
11. Hagen, S., Peck, S., "Out-of-Pile Bundle Temperature Escala-
     tion Under Severe Fuel Damage Conditions," KfK-3568,
     August 1983.
12. McCardell, R. et al., "Severe Fuel Damage Test Series: Severe
     Fuel Damage Scoping Test Quick Look Report," EG&G Idaho, Inc.,
     January 1984.
13. Rogovin, M., "Three Mile Island--A Report to the Commis-
     sioners and to the Public," Nuclear Regulatory Commission
     Special Inquiry Group Report.
14. McCardell, R. et al., "Severe Fuel Damage Test 1-1 Quick Look
     Report," EG&G Idaho, Inc., October 1983.
15. IDCOR Technical Summary Report, "Nuclear Power Plant Response
     to Severe Accident, November 1984.

RECEIVED October 1, 1985

3

Fission Product Behavior

Paul G. Voillequé

Utility Services Department, Science Applications International Corporation,
Idaho Falls, ID 83402-0696

During the TMI-2 accident, radioactive fission products
were released from the fuel to the reactor coolant.
Radionuclides were subsequently transported from the
reactor primary system to the Reactor Building sump and
into the Auxiliary Building. The residual radionuclide
distributions have been studied to gain insight into the
physical and chemical processes operative during and soon
after the accident. Isotopes of (a) the radioactive noble
gases (Kr and Xe), (b) radioiodine, (c) radiocesium, and
(d) radiostrontium are of principal concern. The analyses
of post-accident fluid and solid samples have identified
60--70 percent of the inventories of each of the four
radionuclide categories. Except for the noble gases, the
entire inventories are contained in the plant. The observed
distributions indicate differences between the behavior of
iodine and cesium, either during or after the accident.

The elevated temperatures experienced during the accident at Three
Mile Island Unit 2 (TMI-2) caused the release of radioactive fission
products from the fuel. Radionuclides were transported through the
primary system piping to the Reactor Building and to the Auxiliary
Building. Knowledge of the post-accident distributions of fission
products is important to understanding the release and transport of
radionuclides under accident conditions. This understanding will
improve the capability to assess potential consequences of various
postulated accidents at other reactor facilities.

The purpose of this paper is to summarize our knowledge of the
post-accident distribution of important fission products at TMI-2.
Emphasis is placed on the noble gas radionuclides and upon radio-
isotopes of iodine, cesium, and strontium because of their importance
to accident dose assessment. Measurements of the radionuclide concentration in a medium and a measurement or estimate of its mass or
volume yield the total radionuclide activity present. This is then
expressed, with appropriate correction for radioactive decay, as a
fraction of the radionuclide inventory present at the time of the
accident.

The initial inventories of the radionuclides of greatest
interest are shown in Table I. The inventories were computed using

0097-6156/86/0293-0045$06.00/0
© 1986 American Chemical Society

the ORIGEN2 computer code (1) and the TMI-2 operating history (2).
There are some differences between the inventories estimated using
the ORIGEN2 code and those estimated using other computer codes (3).
However, the discrepancies in calculated inventories are small and
have no effect on the inventory assessments presented here. The
half-lives of these radionuclides (4) are also shown in Table I.
Each of the categories (noble gas, radioiodine, radiocesium, and
radiostrontium) has a long-lived nuclide that can be measured even
several years after the accident.

Table I. Inventories and Half-Lives of Principal Radionuclides

Radionuclide	Initial Core Inventory (Ci)	Half-Life
^{85}Kr	9.7×10^4	10.7 y
^{133}Xe	1.5×10^8	5.25 d
^{129}I	2.2×10^{-3}	1.6×10^7 y
^{131}I	7.0×10^7	8.04 d
^{137}Cs	8.4×10^5	30.17 y
^{90}Sr	7.5×10^5	28.8 y

The only significant releases of radionuclides to the environment
were releases of noble gases. The releases of other radionuclides
to the environment were very small fractions of the initial inventories. In principle, measurements of the residual inventories of
radionuclides in fluids and solids that remain in the plant should
lead to a complete accounting of the initial core inventories shown
above.
The pathways along which radionuclides moved during and after
the accident are discussed in the next section. These transport
routes define the locations of interest in a post-accident inventory
assessment. Following that, a brief description of the sampling and
analyses that have been performed is presented. It cannot begin to
describe the number and magnitude of the problems that have been
faced and overcome in gathering, shipping, and analyzing the samples
from TMI-2. The concerted efforts of many dedicated people have
provided the data upon which this current assessment is based. The
calculated radionuclide inventory distributions are summarized, by
location, for the four radionuclide categories and the implications
of the results are discussed.

Radionuclide Transport Pathways
The pathways for transport of radionuclides from the core into the
Reactor Building and the Auxiliary Building are physically limited
to those provided by the primary system piping. Tracing these
pathways identifies possible locations of the radionuclides released
during the accident.

Radionuclides released from the fuel entered an atmosphere of
steam and hydrogen in the reactor vessel void space that would
normally be filled with liquid coolant. A fraction of the
radioactivity deposited on various surfaces in the reactor vessel,
on the large steel surface area of the plenum located above the
core, and on the interior walls of the primary system piping, the
pressurizer, and the steam generators. After the vessel was
refilled with coolant, the core debris and the deposits on interior
surfaces were in contact with the liquid. There was continued
leakage of coolant into the Reactor Building during the months
following the accident. This leakage provided a mechanism for
delayed transport of activity into the Reactor Building.

A principal route for material transported promptly from the
primary system to the Reactor Building was via the pressurizer and
reactor coolant drain tank. Large amounts of coolant and noble gas
radionuclides followed this path. Once in the Reactor Building, the
radionuclides could be found in the water in the basement, in the
associated sediments, on Reactor Building surfaces, or in the
Reactor Building atmosphere.

Radionuclides were transported to the Auxiliary Building via the
letdown system, which contains filters and demineralizers for purification of the reactor coolant and includes several large tanks in
the Auxiliary Building. The principal radionuclide releases to the
environment were from the tank vents and the Auxiliary Building
ventilation system. The effluent gas was filtered prior to discharge, which removed radionuclides and reduced the releases of
radioiodines and particulate material. Deposition on the interior
surfaces of piping and ventilation ducts as well as on walls and
other surfaces within the building is also a possible removal
mechanism for particulates and radioiodine.

Radionuclide Measurements

Knowledge of the transport pathways described above has guided the
program of measurements to determine the radionuclide inventories at
various locations in the facility. To determine the radionuclide
inventory in the sampled medium, an estimate of the total mass,
volume, or surface area is needed in addition to the measured
radionuclide concentration. Table II contains the locations of
radionuclide concentration measurements in the Reactor Building
(RB). The sampling locations in the left column address residual

Table II.   Reactor Building Fission Product Measurements

Reactor Coolant	RB Atmosphere
Core Debris	RB Sump Liquid and Solid
Plenum Surfaces	RB Interior Surfaces
Lead Screw Surfaces	Reactor Coolant Drain Tank
	RB Purge Discharge

inventories within the primary system, while those on the right side
of Table II deal with other parts of the Reactor Building.

Radionuclide concentration measurements in the Auxiliary Building (AB) have been performed for locations shown in Table III. As
in Table II, locations in the left column of Table III are part of
the reactor coolant or primary system. Sampling locations on the
right side of Table III deal with residual radionuclides outside
that system.

Table III. Auxiliary Building Fission Product Measurements

Coolant Filters and Demineralizers	Effluent Air Discharges Air Filtration Media Effluent Liquid Discharges
AB Coolant Storage Tanks	AB Sump Liquids and Solids

The time sequence of radionuclide measurements is an important
factor in their interpretation. The measurements of reactor coolant
concentrations were initiated promptly (37 hours after the accident)
and continued for a long time period. Other media could not be
sampled until weeks, months, or even years, in the case of the core
debris, had elapsed. There is also evidence (5) that the distribution of iodine between the solid and liquid phases in the Reactor
Building sump changed with time after the accident. Other examples
of the time dependence of radionuclide inventory distributions are
discussed below.

Radionuclide Inventory Distributions

The sampling and analysis difficulties and long time delays prior to
the availability of radionuclide data for some of the locations in
Tables II and III have made assessment of the TMI-2 fission product
distributions an extended process. Even now, six years after the
accident, the picture remains incomplete. The inventory distributions for noble gases, iodine, cesium, and strontium presented in
this paper are based upon the currently available data set. Future
measurements will be needed to complete the picture.

The first category considered is the fraction of the inventory
released to the environment. This category is significant only for
the noble gases. The inventories of other radionuclides were contained within the plant. Attention is then focused on the in-plant
fluids, surfaces, and solids that are known or are likely to contain
major fractions of the inventories of iodine, cesium, and strontium.

Releases to the Environment. The noble gases are the most volatile
of all radionuclides produced in nuclear reactors. The releases to
the environment soon after the accident consisted almost entirely of
noble gases. Approximately 8% of the core inventory of ^{133}Xe was
released from the Auxiliary Building during the first 10 days after
the accident (6). Smaller amounts were released periodically at
later times, but the releases were only small fractions of the
initial discharge.

The next major release of noble gas radionuclides was during the
controlled purge of the Reactor Building about 15 months after the
accident. The long-lived isotope ^{85}Kr was the principal radionuclide
in the purge air. It was estimated that approximately 46% of the
^{85}Kr inventory was discharged under good dispersion conditions
during the Reactor Building purge (7).

Based on the initial release and purge discharge measurements
cited above, the total release of noble gases from the fuel would be
less than 60% of the initial inventory. This is somewhat lower than
the earlier estimate of Bishop et al. (8) who used early Reactor
Building atmosphere sampling results to estimate a fuel release
fraction of 70%. The location of the remaining 30-40% of the inventory is presently unknown. Core debris that experienced relatively
low temperatures during the accident may contain some of the noble
gas activity. A detailed review of the many smaller noble gas
releases that occurred during 1979 and 1980 would improve the
accounting for noble gas radionuclides.

The releases of radioiodines to the environment were very small,
so essentially all of the inventory must remain within the facility.
Measurements of ^{131}I showed that less than 0.00004% of the inventory
was released from the Auxiliary Building (6). At the start of the
Reactor Building purge, about 0.002% of the ^{129}I inventory was in
the Reactor Building atmosphere. Releases of cesium and strontium
to the environment were even smaller than those for iodine.

The minimal radioiodine release was one of the surprising
aspects of the TMI-2 accident (9). In accident consequence assessments, the releases of radioiodines have been assumed to be quite
large for accidents as severe as the one at TMI-2 (10,11). The
lower than expected releases from TMI-2 meant that radiation doses
to members of the public living near the plant were also smaller
than those that have been predicted (12). This has led to a re-
evaluation of accident source terms (13). The potential regulatory
impact of the findings is discussed by Pasedag in these Proceedings
(14).

Major Fluid Reservoirs. Table IV shows the fractions of the radionuclide inventory that have been found in the major liquid volumes
in the Reactor and Auxiliary Buildings. For radioiodines, a range
of estimates is shown. This range reflects discrepancies among
samples collected at different times and variations between data for
^{129}I and ^{131}I. The measured concentrations of the single isotopes
of cesium and strontium were much less variable than those for the
two iodines.

Table IV. Radionuclide Distributions in Major Fluid Reservoirs

	Percentage of Core Inventory[a]		
Location	Iodine[b]	Cesium	Strontium
Reactor Coolant	2 (1-3)	1	1
RB Sump Liquid and Solids	17 (6-30)	41	2
AB Coolant Storage Tanks	3 (2-5)	3	0.09

a. Distribution on 28 August 1979.

b. Values in parentheses show the range of estimates based upon
    samples collected at different times and analyses for different
    radionuclides.

Little fission product activity was found in the sample of
liquid and solids in the reactor coolant drain tank, which is part
of the flow path from the pressurizer to the sump. Although the
mass of solids is somewhat uncertain, this location does not contain
an important fraction of the inventory. The calculated amount of
iodine in the sump is also dependent upon the estimate of mass of
solids present there. Later measurements indicate an increase in
the activity in solids and a decreased amount of iodine in sump
liquid (5).

The distributions in Table IV are referenced to 28 August 1979,
the date the first sample of Reactor Building sump liquid and solids
was collected. Analyses (15, 16, 17) have shown that much of the
iodine, cesium, and strontium activity entered the sump via coolant
leakage many days after the accident. A more recent analysis of
radiocesium behavior (18) has shown the appearance rate in the
coolant to be comparable to laboratory data for leaching of cesium
that had been deposited on stainless steel.

The results show a significant difference between the iodine
and cesium inventories in the Reactor Building sump, while the
amounts transported into the Auxiliary Building tanks were comparable. The ^3H content in those tanks was also about 3 percent
of the ^3H inventory (16).

The lower fraction of the strontium inventory identified is
consistent with the expectation that the fractional release of
strontium from fuel at a given temperature would be much less than
for iodine and cesium (19). The concentration of strontium in the
coolant increased to a relatively constant value soon after the
accident, suggesting an equilibrium between the coolant and a
precipitate (20).

It is reasonable to presume that both iodine and cesium were
deposited on the interior surfaces of the reactor vessel, plenum,
and primary system piping during the period when these spaces were
void of water and at high temperatures. When the primary system
water levels were reestablished, these radionuclides were gradually
leached into the coolant and were subsequently transported into the
reactor building sump. Calculations based on the estimated coolant
leakage rates and measured reactor coolant concentrations suggest
that about 23% of the iodine inventory (16) was transported to the
sump at times later than 1.5 days after the accident. That estimate exceeds the average value in Table IV, but lies within the
range shown in parentheses. Comparable calculations (17) indicated
that approximately 25% of the cesium inventory and nearly 8% of the
strontium inventory was transported via coolant leakage during the
same time period. The cesium and iodine transport estimates are
within the ranges estimated previously (15). The original surface
deposits within the primary system would have been depleted by this
process.

Residual Surface Deposits. Measurements of residual radionuclide
deposition on the leadscrew and plenum surfaces indicate that either
the deposition was small or that most of the deposited radioactivity
has been removed. Radionuclide inventory estimates based on these
surface samples, which were not obtained (21) until years after the
accident, are shown in the upper part of Table V. As is the case
for all of the inventory estimates, the actual samples represent
only small fractions of the total. However, if the samples obtained
are representative, the results in Table V show that very small
fractions of the radionuclide inventories remain on the plenum and
leadscrew surfaces (22, 23). The nature of the residual cesium
deposit on th leadscrew surface has been investigated in detail and
the results are described elsewhere in this volume (24).

Table V. Residual Radionuclide Inventories on Surfaces

Location	Percentage of Core Inventory		
	Iodine	Cesium	Strontium
Plenum Surfaces	0.2	0.05	0.04
Leadscrew Surfaces	0.002	0.001	0.001
RB Interior Surfaces[a]	0.4	0.04	0.002

a. Sample surfaces do not include unpainted concrete surfaces in
the basement.

A gradient in concentrations of all three radionuclides was observed in the data for the leadscrews. The concentrations about 3.2 m above the top of the core were generally 4-50 times higher than those only 0.2 m above the top of the core. For ^{129}I, the concentration on the upper portion of the center leadscrew (H-8) was nearly 3000 times that on the lower section. It is not clear whether the variations in residual activity are due to different initial depositions or to different exposures to coolant after the deposition was complete.

Measurements results for other surfaces within the primary system are not yet available. Laboratory measurements of deposition velocity show that it is a strong but not monotonic function of temperature (25). Detailed temperature estimates would be needed to compute radionuclide deposition on surfaces in various parts of the primary system. Deposition on other surfaces at higher temperatures during the period when the water level was low could have exceeded that in the plenum.

In contrast with the estimates of inventory on surfaces inside the reactor vessel, which are based on recent samples, the first measurement of the Reactor Building surface contamination was made in June of 1979. That indirect estimate was obtained using a collimated gamma spectrometer (26). The first physical sample was obtained at the end of August 1979, and many additional samples were collected during containment entires following the containment purge in 1980. These included samples of opportunity, (e.g., paint chips) collected during containment entries (27) as well as more systematic sampling (5).

Available data suggest that less than one percent of the core inventory of radioiodine resides on Reactor Building surfaces. Even smaller fractions of the cesium and strontium inventories were found on the same surfaces. However, in the lower level of the Reactor Building there is a section of the concrete wall that is unpainted. Samples have not been obtained from this area because of the high radiation fields in the basement. Because the unpainted wall surface is near the level of release of coolant into the Reactor Building and because the relatively porous concrete surface would retain more activity than painted surfaces, it is probable that surface radionuclide concentrations in that area exceed those measured at higher levels in the containment. Radionuclide attachment to the unpainted concrete could have occurred by deposition of airborne material early in the course of the accident or by contact with contaminated water at later times.

Reactor Building Atmosphere. In accident consequence calculations, radionuclides in the Reactor Building atmosphere are commonly assumed to be the principal potential source of release to the environment (28). Therefore, measurements of airborne radionuclides at TMI-2, particularly of radioiodines, were of great interest. The first containment air sample indicated the largest amount of airborne radioiodine. The inventory fraction computed using the average results of two analyses of that sample, collected at 75 hours, was 0.005% (16).

The containment sprays, which operated approximately 10 hours after the accident began, removed activity from the containment atmosphere so even the first air sample measured only the residual activity. The Reactor Building surface concentration measurements cited above were used to make an estimate of the maximum amount of iodine that was airborne in the Reactor Building (16). It was estimated that at most 0.2% of the radioiodine inventory may have been in the containment atmosphere following the accident. This estimate is speculative because of uncertainties about the appropriate deposition velocities for iodine species entering into the Reactor Building atmosphere.

Letdown Demineralizers. The two letdown demineralizers, termed "Demin-A" and "Demin-B", were operated for varying times following the release of fission products from the fuel. The flows of contaminated coolant through the demineralizers are not well known. Because of the high radiation fields around them, the remains of the demineralizer resins were not sampled until 1983 (29).

One analysis for iodine has been reported for the samples of resin and water taken from the letdown demineralizers. The measured concentration in the dry Demin-A resin was 10^{-4} g ^{129}I/g resin. No measurements were made of ^{129}I in the two Demin-B resin or the two Demin-B liquid samples. The total mass of the two resin beds is estimated to be 7.6×10^5 g. If the measured concentration is applicable to the total mass, approximately 6% of the ^{129}I inventory is in the demineralizers (30).

Measurements of ^{137}Cs in the two resin and liquid samples of Demin-B averaged 1.4×10^4 and 2.1×10^3 μCi/g, respectively. The ^{137}Cs concentration in the Demin-A solids sample was only 220 μCi/g. The estimated fraction of the ^{137}Cs inventory in the two demineralizers is 0.75%. Note that the difference in measured ^{137}Cs concentrations in the Demin-A and Demin-B resin samples may mean that the assumption (made above) of equal concentrations made for ^{129}I is incorrect.

The average concentrations of ^{90}Sr in the Demin-B resin sample were also greater than those in the resin from Demin-A. The concentration averaged 690 μCi/g for the two Demin-B samples and was 200 μCi/g in the sample from Demin-A. It is estimated that 0.05% of the original ^{90}Sr inventory is in the liquid and resins in the demineralizer tanks.

Plans for decontaminating the demineralizers are discussed in this proceedings (31). It is possible that samples from fluids used to flush the demineralizers may be analyzed for ^{129}I. If so, the uncertainty about the fractional radioiodine inventory in the demineralizers may be reduced.

Core Debris. Measurement of the radionuclide content of the core debris is, of course, one of the most important sources of data on the behavior of fission products during the accident at TMI-2. The earliest assessments necessarily relied on the radionuclide concentrations measured in the Reactor Building sump and other locations that were accessible. However, in the past year, measurements of the radionuclide concentrations in samples of the core debris have been completed. This section employs reported analytical results to estimate the fractional inventories of iodine, cesium, and strontium in the core debris.

The collection and analysis of the five initial core debris samples was a major step forward in the overall analysis of fission product behavior at TMI-2. The sampling procedures, sample properties, and analytical techniques have been described in this volume (32) and elsewhere (33,34) by the staff of EG&G Idaho, Inc. In addition to analyses for the fission products considered in this paper, concentrations of other radionuclides and of stable elements are being measured. Those results will provide new dimensions to the analysis of the TMI-2 accident.

The results used in this paper are those reported for iodine, cesium, and strontium concentrations in aliquots of the five samples of core debris that have been collected (33,35). The results for cesium in larger composites differ somewhat from the results for the aliquots. Analyses for strontium in the composites are underway; however, analyses for iodine are not planned for the composites. The small aliquots are the only samples for which measurement results for all three radionuclides are available. Results for composite samples are described elsewhere in this volume (32). Until differences between the results are understood or more analytical results become available, estimated cesium inventories of the radionuclides in core debris will differ because the concentration estimates are not the same.

The masses of the five core debris samples and of the aliquots of those samples that have been analyzed for all three radionuclides are shown in Table VI. The total sample mass is about 0.5 kg. The fissile content of each aliquot, also shown in Table VI, indicates that the material is principally fuel, although smaller subsamples that are definitely cladding have also been identified (33).

Table VI. Description of Core Debris Samples and Aliquots
of the Samples

Debris Bed Samples	Sample Mass (g)	Aliquot Mass (g)	Aliquot Fissile Content (%)
Center Location H-8			
Surface	69	0.56	2.4
56 cm Deep	148	0.62	2.4
Axial Location E-9			
Surface	17	0.081	--
5 cm Deep	90	0.73	2.6
56 cm Deep	138	5.6	2.9

All of these samples were collected from the bed of loose debris which lies atop a firm layer of consolidated or resolidified core material approximately 2 m below the plenum. This solid debris layer has not been sampled yet, nor has the loose and/or resolidified debris that is present beneath the core support plate (11). Concentrations of fission products in those components of the core debris may well differ from those in the samples analyzed to date. Future sampling and analyses will be performed as the defueling of TMI-2 proceeds. The planned future activities are described elsewhere in this volume (36).

The concentrations measured in the aliquots of the five core debris samples were used to estimate the iodine, cesium, and strontium radionuclide inventories in core debris. The assumption was made that the samples are representative of the approximately 120 MT of debris. As noted above, the debris bed is not uniform in character, so this assumption may be invalid. Table VII shows the inventory estimates based on analytical results for the aliquots for each core debris sample.

Table VII.   Estimated Radionuclide Inventories in Core Debris

	Percentage of Core Inventory[a]		
Debris Bed Samples	Iodine	Cesium	Strontium
Center Location H-8			
Surface	43	7.5	58
56 cm Deep	26	10	76
Axial Location E-9			
Surface	8.3	14	61
5 cm Deep	37	20	68
56 cm Deep	31	19	14
Arithmetic Mean	29	14	66[b]
Std. Dev. of Mean	7	3	4[b]

a. Calculated using analytical results for aliquots described in Table VI.

b. Based upon the first four samples listed.

The fractional inventory estimates for iodine and strontium exhibit considerable variation; the ranges of the estimates are 8-43 percent and 14-76 percent, respectively. The variability in the fractional cesium inventory estimates is smaller, with estimates ranging from 8-20 percent.

The ratio of the fractional iodine inventory to the fractional cesium inventory varies from 0.6 to 6, but ratios for four of the five samples exceed 1.6. The arithmetic mean value for iodine in the core debris is double that for cesium. This partly explains the observed difference between the fractional iodine and cesium inventories in the Reactor Building sump. However, it is not known whether the iodine release from the fuel was lower than that for cesium or whether released iodine was later bound to the fuel debris.

The estimated strontium inventory fractions are smaller than expected. The value of 14 percent for the last sample seems very unusual and it was not used in the calculation of the mean inventory fraction. Data suggest that approximately one-third of the strontium was released from the fuel. This implies much higher fuel temperatures than were estimated in previous thermal hydraulic calculations (37). Recent thermal hydraulic evaluations, presented in this volume (11), also suggest higher temperatures and a longer period at elevated temperatures.

The fraction of the strontium inventory that was apparently released from the fuel was not found in the sump or other plant samples. This suggests that it was precipitated in the reactor vessel or deposited on surfaces, either in the vessel or on the concrete in the Reactor Building basement. A combination of both mechanisms is perhaps most reasonable. Calculations using measured coolant concentrations and estimated leakage rates suggested that about 8% of the strontium inventory entered the sump in the first five months after the accident (17); however, only about 2% of the inventory was found there (Table IV). Deposition on the basement walls is one mechanism that could remove strontium from the sump liquid and solids.

Conclusions

Measurements of fission product concentrations in a wide range of samples following the accident at TMI-2 have been performed by many individuals and groups in the face of great obstacles. The measurements showed that, except for about 8 percent of the noble gas radionuclides, accidental releases to the environment were extremely small. Except for decreases due to radioactive decay, the inventories of iodine, cesium, and strontium radionuclides will be found within the plant. Measurements to date have been used to locate about 60 percent of the iodine and cesium inventories, and about 70 percent of the strontium inventory. These estimates are approximate and depend upon simplifying assumptions that require validation by future measurements.

The recently obtained data for core debris indicate fuel temperatures were much higher than those estimated previously. This result is generally consistent with the evolving picture of the physical processes that occurred during the first hours of the accident.

The observed distributions of iodine and cesium are different. Approximately 29% of the iodine is associated with core debris, compared with 14% of the cesium. Much more of the cesium inventory

(41%) was found in the sump; the mean estimate for iodine is 17% with a range of 6-30%. The initial releases of these two nuclides from the fuel were expected to be comparable (19,38). If that expectation was realized, then there were differences in deposition within the primary system immediately after release and/or in the removal and transport processes that occurred after the vessel was refilled with coolant.

Literature Cited

1. Croff, A. G., Nuclear Technology 1983, 62, 335.
2. Canada, R. G., "NSAC EPRI ORIGEN Code Calculation of TMI-2 Fission Product Inventory," Report No. R-80-012 Technology for Energy Corporation: Knoxville, TN, 1980.
3. Davis, R. J., Tonkay, D. W., Vissing, E. A., Nguyen, T. D., Shawn, L. W., and Goldman, M. I., "Radionuclide Mass Balance for the TMI-2 Accident: Data Through 1979 and Preliminary Assessment uf Uncertainties," Report NUS-4432, NUS Corporation: Gaithersburg, MD, 1983.
4. Lederer, C. M., and Shirley, V. S., Eds.; "Table of Isotopes," 7th Ed., Wiley-Interscience: New York, 1978.
5. McIsaac, C. V. and Keefer, D.G., "TMI-2 Reactor Building Source Term Measurements: Surfaces and Basement Water and Sediment," DOE Report GEND-042, EG&G Idaho, Inc., Idaho Falls, ID, 1984.
6. Pickard, Lowe, and Garrick, Inc., "Assessment of Offsite Radiation Doses from the Three-Mile Island Unit 2 Accident," Report TDR-TMI-116, Middletown, PA, 1979.
7. Kripps, L. J., "TMI-2 Reactor Building Purge--[85]Kr Venting," DOE Report GEND-013, EG&G Idaho, Inc., Idaho Falls, ID, 1981.
8. Bishop, W. N., Nitti, D. A., Jacob, N. P., and Daniel, J. A., "Fission Product Release from the Fuel Following the TMI-2 Accident," in Proceedings of the ANS/ENS Topical Meeting on Thermal Reactor Safety, Conference Report CONF-800403, 1980.
9. Stratton, W. R., Malinauskas, A. P., and Campbell, D. O., Letter to NRC Chairman Ahearne, August 14, 1980.
10. Thomas, G. R., "Description of the TMI-2 Accident," these Proceedings.
11. Tolman, E. L., Allison, C. M., Davis, C. B., and Polkinghorne, S. T., "Thermal Hydraulic Aspects of the TMI Accident," these Proceedings.
12. Nuclear Regulatory Commission, "Reactor Safety Study, An Assessment of Accident Risks in U.S. Commercial Nuclear Power Plants," NRC Report WASH-1400, NUREG-75/014, Nuclear Regulatory Commission, Washington, D.C., 1975.
13. American Nuclear Society, "Report of the Special Committee on Source Terms," Unnumbered Report, American Nuclear Society, La Grange Park, IL, 1984.
14. Pasedag, W. F., "The Impact of TMI on Future Licensing," these Proceedings.
15. Pelletier, C. A., Thomas, C. D. Jr., Ritzman, R. L., and Tooper, F., "Iodine-131 Behavior During the TMI-2 Accident," EPRI Report NSAC-30, Electric Power Research Institute, Palo Alto, CA, 1981.

16. Pelletier, C. A., Voilleque, P. G., Thomas, C. D. Jr., Daniel, J. A., Schlomer, E. A., and Noyce, J. R., "Preliminary Radio-iodine Source Term and Inventory Assessment for TMI-2," DOE Report GEND-028, Science Applications, Inc., Rockville, MD, 1983.
17. Voilleque, P. G., Noyce, J. R., and Pelletier, C. A., "Estimated Source Terms for Radionuclides and Suspended Particulates During TMI-2 Defueling Operations, Phase II," DOE Report GEND-INF-019, Science Applications, Inc., Rockville, MD, 1983.
18. Lorenz, R. A., Campbell, D. O., Malinauskas, A. P., Collins, E. D., Lowrie, R. S., Culberson, O. L., and Collins, J. L., "TMI-2 Radiocesium Behavior," in Proceedings of the ANS Topical Meeting on Fission Product Behavior and Source Term Research, American Nuclear Society, La Grange Park, IL, 1985.
19. Nuclear Regulatory Commission, "Technical Bases for Estimating Fission Product Behavior During LWR Accidents," NRC Report NUREG-0772, Nuclear Regulatory Commission, Washington, D.C., 1981.
20. Malinauskas, A. P. and Campbell, D. O., Oak Ridge National Laboratory, Personal communications, 1981.
21. Carlson, J. O., "Implementation Plan for TMI-2 H8, B8, and E9 Leadscrew Examination," Draft DOE Report EGG-TMI-0002, EG&G Idaho, Inc., Idaho Falls, ID, 1983.
22. Vinjamuri, K., "TMI-2 Leadscrew Exam," Proceedings of the ANS Topical Meeting on Fission Product Behavior and Source Term Research, American Nuclear Society, La Grange Park, IL, 1985.
23. Vinjamuri, K., Akers, D. W., and Hobbins, R. R., "Examination of H8 and B8 Leadscrews from Three Mile Island Unit 2 (TMI-2)," Draft DOE Report EGG-TMI-6685, EG&G Idaho, Inc., Idaho Falls, ID, 1984.
24. Baston, V. F. and Hofstetter, K. J., "Adherent Activity on TMI-2 Internal Surfaces," these Proceedings.
25. Nicolosi, S. L. and Buybutt, P., "Vapor Deposition Velocity Measurements and Correlations for I_2 and CsI," NRC Report NUREG/CR-2713, Battelle Columbus Laboratories, Columbus, OH, 1982.
26. Bechtel Power Corporation, "Planning Study for Containment Entry and Decontamination," Unnumbered report, Bechtel Power Corporation, Gaithersburg, MD, 1979.
27. Science Applications, Inc., "Analysis of TMI-2 Paint Chip Samples," Unnumbered report, Science Applications, Inc., Rockville, MD, 1981.
28. Nuclear Regulatory Commission, "Assumptions Used for Evaluating the Potential Radiological Consequences of a Loss of Coolant Accident for Pressurized Water Reactors," Regulatory Guide 1.4, Revision 2, Nuclear Regulatory Commission, Washington, D.C., 1974.
29. Thompson, J. D. and Osterhoudt, T. R., "TMI-2 Purification Demineralizer Resin Study," DOE Report GEND-INF-013, EG&G Idaho, Inc., Idaho Falls, ID, 1984.
30. Cox, T. E. and Quinn, G. J., "Makeup and Purification Demineralizers at TMI-2," Draft DOE Report, EG&G Idaho, Inc., Idaho Falls, ID, 1984.

31. Bond, W. D., King, L. J., Knauer, J. B., Hofstetter, K. J., and Thompson, J. D., "Cleanup of TMI-2 Demineralizer Resins," these proceedings.
32. Akers, D. W., "TMI-2 Core Debris Chemistry and Radionuclide Behavior," these Proceedings.
33. Akers, D. W. and Cook, B. A., "Draft Preliminary Report: TMI-2 Core Debris Grab Samples - Analysis of First Group of Samples," Draft DOE Report EGG-TMI-6630, EG&G Idaho, Inc., Idaho Falls, ID, 1984.
34. Cook, B. A., Nitschke, R. L., and Akers, D. W., "TMI-2 Core Debris Analytical Methods and Results," in Proceedings of the ANS Topical Meeting on Fission Product Behavior and Source Term Research, American Nuclear Society, La Grange Park, IL, 1985.
35. Akers, D. W., "Preliminary ^{129}I and ^{90}Sr Analysis Results and Correlations," Letter report, EG&G Idaho, Inc., Idaho Falls, ID, 1984.
36. Eidam, G. R. and DeVine, J. C. Jr., "TMI-2 Current Status and Plans," these Proceedings.
37. Ireland, J. R., Wehner, T. R., and Kirchner, W. L., Nuclear Safety 1981, 22, 583.
38. Paquette, J. and Wren, D. J., "Iodine Chemistry," these Proceedings.

RECEIVED July 16, 1985

4

Assessment of Thermal Damage to Polymeric Materials by Hydrogen Deflagration in the Reactor Building

N. J. Alvares

Lawrence Livermore National Laboratory, Livermore, CA 94550

Thermal damage to susceptible material
in accessible regions of the reactor
building was distributed in nonuniform
patterns. No clear explanation for non-
uniformity was found in examined evidence,
e.g., burned materials were adjacent to
materials that appear similar but were not
burned. Because these items were in
proximity to vertical openings that extend
the height of the reactor building, we assume
the unburned materials preferentially absorbed
water vapor during periods of high, local
steam concentration. Simple hydrogen-fire-
exposure tests and heat transfer calculations
duplicate the degree of damage found on
inspected materials from the containment
building. These data support estimated
8% pre-fire hydrogen concentration predictions based on various hydrogen production
mechanisms.

About 10 hours after the 28 March 1979 loss-of-coolant accident
began at the Three Mile Island Unit 2 Reactor Building, a hydrogen
deflagration of undetermined extent occurred inside the reactor
building. Hydrogen was generated as a result of reaction between
zirconium nuclear fuel rod cladding and steam produced as the
reactor core was uncovered. Figures 1 through 4 are extracted from
a variety of resources (indicated on figure), and they summarize
the conditions and evidence of hydrogen release. Figure 1 is a
schematic of the reactor coolant system showing the path of
hydrogen release. Measurements of background activity increase
(Figure 2) show the release occurred about 2.5 hours past turbine
trip. Henrie and Postma (3) as well as Zalosh and others (1)
estimate hydrogen accumulation in the core by a variety of means
(Figure 3):

0097-6156/86/0293-0060$07.75/0
© 1986 American Chemical Society

o   Timing of projected hydrogen generation in the core;
o   Timing the pressure relief valve opening periods;
o   Pressure changes in the reactor coolant system;
o   Calculations of hydrogen mass burned; and,
o   Measurements of post-burn hydrogen concentration.

Figure 4 shows the reactor-building pressure record starting from
the time of reactor trip to well after the combustion produced
pressure pulse.

Interviews with "on duty" plant personnel indicate they did
not perceive that the "thud" they heard was caused by a hydrogen
deflagration in the reactor building. Moreover, paucity of easily
observable damage delayed recognition that a hydrogen burn did
occur for about two days. Ignition of the hydrogen-and-air mixture
release after the breach of the reactor coolant drain-tank (RCDT)
rupture disk resulted in nominal thermal and overpressure damage to
susceptible materials in all accessible regions of TMI-2.
Initiation of burn and subsequent termination of induced fires are
indicated by data from a variety of pressure and temperature
sensors located throughout the containment volume.

Activation of the building spray system is defined by
inflection and increase in the negative slope of interior-pressure-
reduction curves (Figure 5). Also indicated is a pressure increase
of about 28 psig achieved in a period of about 12 s.(3)
Experimental confirmation of the pressure response of hydrogen
combustion in constant volume chambers is indicated in Figure 6.
Note that the act of causing turbulent conditions in the test
chamber causes greater pressure rise at lower hydrogen
concentration.

The hydrogen-in-air concentration [H_2] was estimated to be
approximately 6 to 8%. At this concentration range, propagation of
flame is possible upward and horizontally in quiescent conditions,
but not downward. Figure 7 shows how laminar burning velocity
varies with hydrogen concentration in air. Directed arrows at the
lean and rich regions of hydrogen concentration indicate allowed
flame-spread propagation vectors. This effect occurs because of
competition between fundamental flame speed and buoyancy induced by
reactants temperature rise. Figure 8 shows an example of lean limit
propagation for methane-air-nitrogen mixtures. This illustrates
the effect expected in hydrogen concentrations less than 8% in air.
However, turbulent conditions, established circulation patterns,
and the ambient absolute humidity of the mixture can perturb propagation patterns in ways that are only qualitatively understood.(4-5)
Assuming uniform mixing of 8% hydrogen-in-air concentration and
induction of adequate turbulence in internal circulation flows,
average flame speeds of 5 m/s (16 ft/s) are possible -- even in the
presence of saturated steam environments.(6)

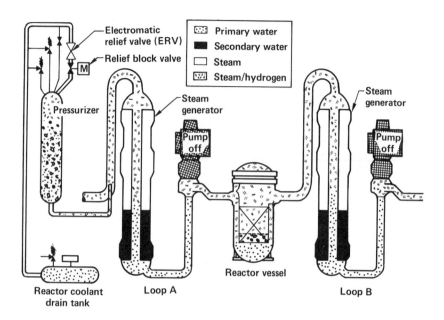

Figure 1. Reactor coolant system schematic showing hydrogen release path through reactor coolant drank tank. Reproduced from Ref. 1.

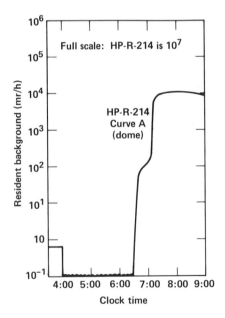

Figure 2. Background activity as measured by sensor in dome. Reproduced from Ref. 2.

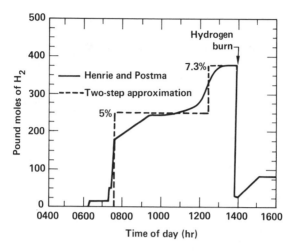

Figure 3. Hydrogen production estimates based on analysis of pre- and post-burn core and reactor building indicators. Reproduced from Ref. 3.

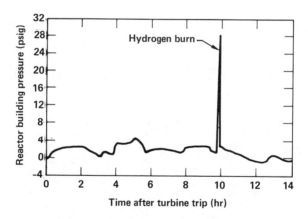

Figure 4. Reactor building pressure record. Reproduced from Ref. 1.

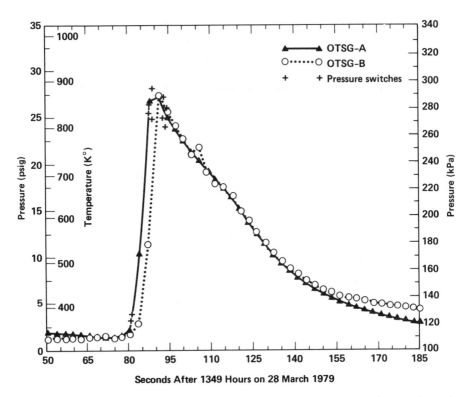

Figure 5. Pressures recorded during the burn from OTSG (once-through steam generator) pressure transmitters and pressure switch actuation times. (Reproduced from Ref. 1.) Corresponding average temperature via procedure described in Ref. 1. added to psig scale.

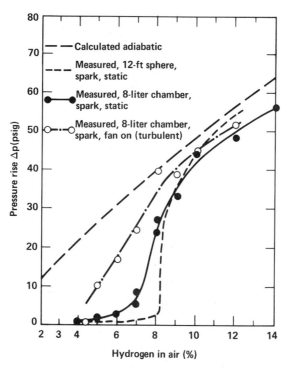

Figure 6.   Pressure rise (p) produced by burning hydrogen ignited in constant volume chambers.   Reproduced from Ref. 4.

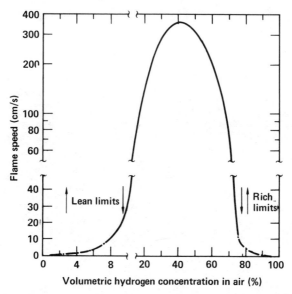

Figure 7.   Flame speed of varying hydrogen in air mixtures. Reproduced from Ref. 4.

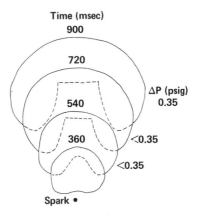

Figure 8.   Lean methane and nitrogen in air mixture flame propagation patterns (6.9% CH_4 – 27.3% N_2 65.8% Air).   Reproduced from Ref. 4.

A cross section of the reactor building (Figure 9) and plan
view of the main (347-ft) operations level (Figure 10) show the
regions of thermal and burn damage. Given that few operational
ignition sources were available in the reactor building above the
305-ft level, the time delay to achieve peak overpressure is consistent with an ignition location in the basement. Potential electrical shorting of electrical control systems caused by basement water
spillage and the frequency of steam and hydrogen release from the
reactor coolant drain-tank pressure-release system supports this
assumption.

Thermal damage to fine fuels (fine fuel is defined as a
flammable material with high surface-to-volume ratio) indicates
general exposure of all susceptible interior surfaces to fire with
the exception of random materials including fabric ties of unknown
composition, 2 x 4 framing lumber on both the 305-ft and 347-ft
levels, and various polymeric materials. These unburned items are
evident in photographic and video surveys, and were visually
reconfirmed by various entry participants. This pattern is
reported in several preliminary reports.(7-8) Possible mechanisms
to prevent thermal damage to these items include:

o   Preferential absorption of water from saturated
        atmosphere, requiring greater thermal exposure to
        produce thermal damage.

o   Direct exposure to high-concentration steam and water
        vapor, requiring greater thermal exposure to produce
        thermal damage.

o   Shielding from thermal radiation by position or geometric
        obscuration.

o   Shielding from the expanding flame front or convectively
        driven hot gases by physical obstruction.

Although photographic surveys of internal reactor building
vistas, ensembles, items, and surfaces were abundant (approx 600
photos from 29 entries), clarity of the burn detail in most
photographs was not adequate for diagnostic purposes. However, the
extent of thermal damage was defined (Figures 9 and 10) as regions
where thermally degraded materials were located, photographed, and,
in some cases, extracted from the reactor building for further
examination.

Ignition of a uniformly distributed near-lower-limit mixture
of hydrogen in air, spreading from basement ignition sources to the
top of the reactor building dome by turbulent propagation modes,
occurred in the time period indicated in Figure 5. The flame front
would have been at an adiabatic flame temperature of about 700°C to
800°C (approximately 1000°K), as shown in Figure 11.

Exact paths of flame propagation are undefined. Because of
the low hydrogen concentration, preferential flame spread was
upward in quiescent atmosphere; however, air motion produced by
reactor building coolers, steam/hydrogen release from the rupture
disk line of the RCDT and flow distortion around obstructions
caused turbulent flow conditions which greatly modify flame spread

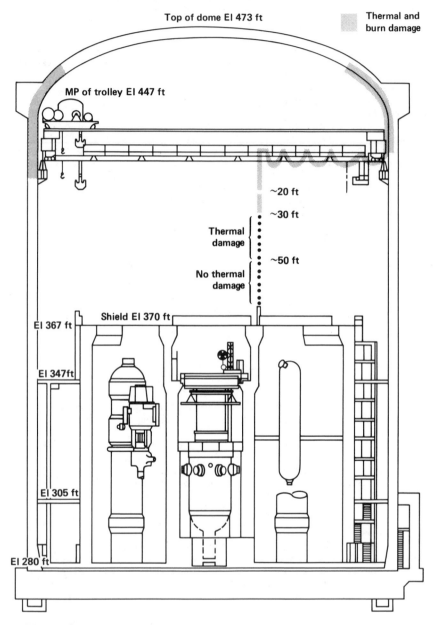

Figure 9.   Cross section of TMI-2 reactor containment building.
Reproduced from Ref. 7.

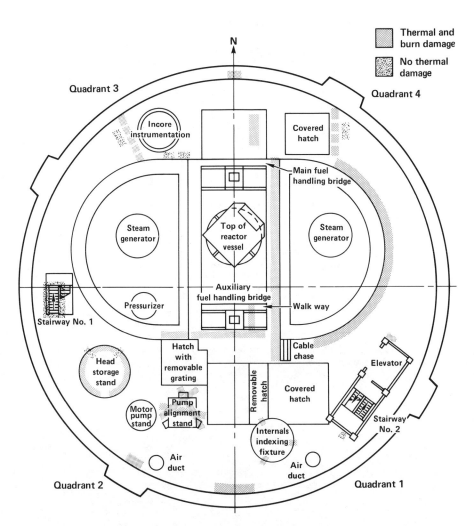

Figure 10. Thermal and burn damage and potential overpressure on the 347-ft level. Reproduced from Refs. 7 and 8.

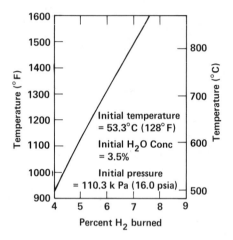

Figure 11.   Temperature rise produced by combustion of pre-mixed hydrogen and air.   Reproduced from Ref. 3.

rates. The source of major hydrogen release was located near the open stairway on the undersurface of the 305-ft level plane (Figure 12). Henrie and Postma (3) conclude that the primary path for entry of the hydrogen-and-steam mixture to the total reactor building above the basement (282 ft) level was through this stairwell. How these gases from the rupture disk line interacted with total ventilation patterns is not known. This may be a moot point since, by the time ignition occurred, hydrogen in the reactor building was substantially mixed.

Identification of a specific ignition source is not possible from available documentation; however, two potential basement source-types are considered. (1) Several circuit boxes, instrument racks, meters, and control components were at various locations around D-rings and containment walls at undefined heights above the basement floor. Failure of circuit components may have been caused by immersion in water. (2) Plant operators who control core and reactor building conditions may have produced ignition arcs from control components perturbed by thermal or mechanical effects of reactor excursion.(2) The inner perimeter of the reactor building basement had no obstructions to block or blind flow of gases outside of the D-ring. However, there were constrictions that could temporarily horizontal hydrogen mixing in the basement.(1) Approximately 10% of the cooled gases from the cooling system plenum (25,000 ft^3/min) was distributed to the basement (outside of the D-ring) through committed ducting. The only exit paths for these gases were the 4-in. seismic gaps, a space that physically separates each floor level from the reactor building, many pipe penetrations and the open stairwell that extended from the basement space to the 347-level without barrier. (A recently identified path for hydrogen release is the in-core instrumentation cable chase which provides a large open area between the basement and the 305-ft level.(12)) If ignition occurred at sources away from the open stairwell, the preferred flame propagation would be upward through the seismic gap, and above the 305-ft level, through the grating in the 347-ft level floor. Horizontal spread would occur, but at a slower rate, even during turbulent propagation conditions. Ample evidence exists on the 347-ft level to confirm flame propagation through the seismic gap regions and the floor grating.

At the peak pressure rise of about 28 psig during the hydrogen burn, the adiabatic temperature rise during combustion of 6 to 8% hydrogen-in-air mixture is about 1000°K. At this temperature, calculated exposure radiative and convective flux ($\dot{q}_t$) from an optically thick combustion plume is

$$2.2 \text{ W/cm}^2 < \dot{q}_t < 4.5 \text{ W/cm}^2$$

This range is approximate because we assume values for combustion zone emittance (ε) at limits of the range $0.2 < \varepsilon < 0.8$. It is quite possible that ε could be larger for optically thick hydrogen combustion zones.(9) Figure 13 compares radiant emission from methane/air fires at various plume diameters. As zone volume increases, the product of emittance times absolute temperature correspondly increases. Since plume temperature is essentially constant, their flame and hot gas emittance is shown to be directly

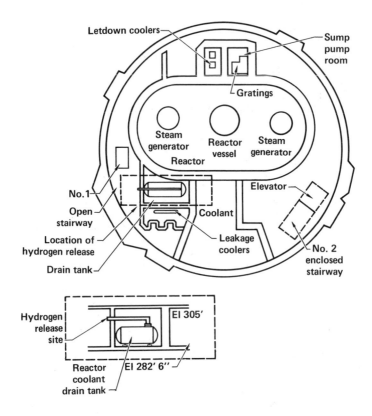

Figure 12. Schematic of reactor basement geometry showing relation of reactor coolant drain tank to relative gas distribution patterns. Reproduced from Ref. 1.

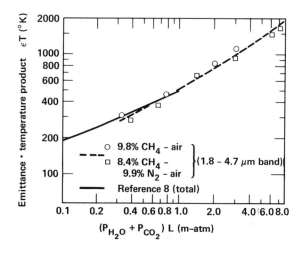

Figure 13. Emittance vs. fire depth. Reproduced from Ref. 9.

proportional to gas volume. Heat transfer coefficient for minimum
and maximum convective heat transfer is based on gas velocity
(u_g) at the limits of the range:

$3 \text{ m/s} < u_g < 12 \text{ m/s}$

Examination of TMI Materials

To estimate the intensity of thermal exposure to damaged materials
and to analyze thermal damage patterns, it is necessary to examine
their condition and to determine their composition. Photographic
evidence is inadequate for such appraisal. We examined materials
removed from the reactor building, and recommended removal of
additional materials for analysis. We examined the following
materials (available July 1983):

Level 305	Level 347	Polar Crane
Polypropylene bucket	Plywood board	Fire extinguisher
	Wood from tool box	Hypalon polar crane pendant jacket control box
	Two radiation signs, probably polyethylene	
	Hemp and polypropylene rope	
	Catalog remains	
	Telephone and associated wire	

These materials retain residual radioactive contamination.
Consequently, all examinations were performed under radiologically-
safe conditions. Chemical or physical analytical procedures could
only be done with contaminated or easily decontaminated instruments.
We were unable to locate expendable diagnostic equipment; therefore,
our examination of extracted materials was limited to detailed
photography and macroscopic observations.
    Figure 14 shows photographs of plywood on the reactor building
south wall and remains of an instruction or maintenance manual
located on the reactor building north wall, both ignited by fire
propagation through the seismic gap and/or radiant exposure from
combustion gases in reactor building free volume. In Figure 14(a)
note the wires along the wall also exhibit burn trauma.
Figures 14(c) and 14(d) show the front and rear surface of the
plywood panel after it was extracted from the south wall of the
reactor building, over the seismic gap. Both sides are charred, as
are edges and holes through which wire ties penetrate. Surface
char condition indicates the panel ignited to flaming combustion
for a short period before self-extinguishing or being quenched by
the reactor spray system. Regardless of the ignition source
location, it is apparent that a hydrogen-and-air flame front

(a)   (b)

(c)   (d)

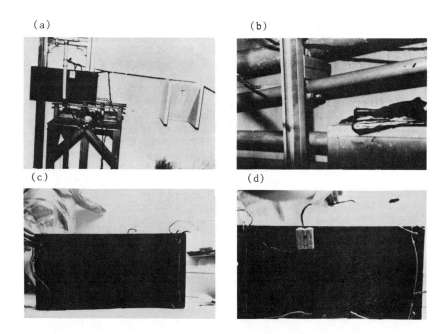

Figure 14. Hydrogen-burned in-containment materials. (a) Bell
telephone at TMI; (b) Charred manual on electrical box; (c)
Plywood panel (back); (d) Plywood panel (front).

traversed most of the reactor building volume above (and probably
below) the 305-ft level. Duration of this propagation was about
12 s. Slow temperature decay before operation of the building
spray system ensured thermal exposure to combustible or thermally
sensitive surfaces was sufficient to produce thermal damage and/or
ignition of these materials, particularly in regions where volume
of the combustion plume was optically thick.

The pendant and festoon for the polar crane possibly received
the most intense energy exposure. The covering of this cable is
Du Pont Hypalon and ethlylene propylene rubber. Figures 15(a) and
15(b) show the lower polar crane pendant, and upper polar crane
pendant and festoon along the "A" girder of the crane.
Figures 15(c) and 15(d) show the relative thermal damage of cable
sections extracted from the reactor building. (This examination
was conducted at Sandia Laboratories, Albuquerque, in cooperation
with Mr. Ralph Trujillo, Project Manager for the cable integrity
project for TMI-2 reactor building electrical circuits.(10)) A
detailed description of thermal damage on each section is contained
in Figure 16, along with a curve showing β/γ activity along the
pendant cable. Figures 15 and 16 show that all sections received
thermal exposure, including those coiled on the D-ring catwalk.
The degree of thermal degradation decreased from the polar-crane
level to the D-ring top, and, in fact, was only apparent on the
bottom pieces where cuts in insulation projected free surfaces of
poor heat transfer. Thermal degration is also apparent on light
lenses of the pendant control box.

Maximum thermal damage occurs in the region from 6 to 10 ft
below the polar crane girder (from about the 440-ft level down to
about the 406-ft level). This region shows locally high β/γ
activity, which may correlate to physical absorption by porous,
charred insulation. Thermal damage is severe and circumferentially
equal in this region. Char depth on the polymer surface averages
1 to 2 mm.

From the 406-ft level to the top of the D-ring (the 367-ft
level), thermal damage is progressively less and becomes more directional, i.e, half of the insulation circumference exhibited a
heavier degree of damage, ranging from char at the 406-ft level to
no perceptible insulation degradation just above the D-ring plane.
The pattern of asymmetric thermal damage along the pendant below
the 406-ft level (a distance approximately 14 ft below polar crane
girder bottom) indicates exposure from a westerly direction.(10)
The extent of thermal damage to other available polymers at about
the 347-ft level indicates intense thermal exposure in southerly
areas of the reactor building. Morever, since all containment
gases above the 347-ft level were convected to the air-cooler intake
plenums in the southern sector just below the 347-ft level, some
preferential ventilation pattern may have influenced fire propagation path. However, because of fewer thermally susceptible
materials in the north reactor building regions, we cannot
confidently compare the south and north experience to define
sources of non-uniform heat flux. Had there been either minimal
thermal experience or other patterns in susceptible polymers in any
other region, we may have had better opportunity to define fire
geometry. One cause for asymmetry of the burn pattern below the
406-ft level can be conjectured: the cable at this height was

(a)   (b)

(c)   (d)

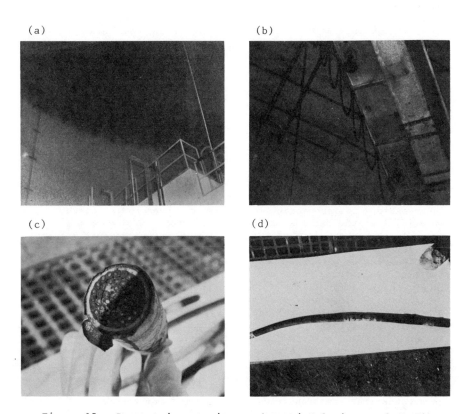

Figure 15. In-containment views and sectional pieces of the
polar crane pendant: (a) Job crane; D-ring A is in lower right;
(b) Girder A of the polar crane; (c) North side of cable is ash;
plastic tape is charred all around; no degradation under the
tape; (d) Half of circumference is ash, half char (ash is gray;
char, black).

exposed to radiation and convection resulting from hydrogen
combustion originating from one side (logically the southwest side)
of the reactor building. The exposed surface would sustain thermal
damage more readily than the shadowed surface, thus producing the
observed pattern.

Photographic documentation of thermal damage patterns
sustained by items removed from the TMI-2 reactor building revealed
a variety of responses from different materials located in the same
general area, e.g., materials around the telephone on the south
reactor building wall of the 347-ft level show quite a different
response relative to material composition.

Thermal Measurements on Exemplar Materials

To augment this analysis, we located exemplar materials generally
similar to those removed from the reactor building. Response
properties of the exemplar materials were measured in a thermal
gravimetric analyzer (TGA) to ascertain the temperature range of
thermal degradation and weight-loss rates. Figure 17 shows TGA
patterns for ABS, acrylonitrile butadiene styrene, a standard
material from the National Bureau of Standards (NBS) used as a
control for smoke tests. ABS is similar to telephone body material.

Thermograms are obtained by isothermally heating milligram-
sized samples of materials, supported on a micro balance, at a
constant temperature rate. Weight loss with temperature indicates
thermal degradation mode and mechanism. The temperature range of
maximum weight loss indicates critical conditions for producing
potentially ignitable pyrolyzates. Figure 17 shows that NBS-ABS
flammable pyrolyzates are produced in the temperature range of 370°
to 500°C, leaving about 20% inert materials as residue. These
pyrolyzates are flammable which, with an external ignition source,
will ignite within this range.

The temperature corresponding to the median of weight loss
during the first major weight-loss experience in any polymer can be
used to estimate the condition where the rate of thermal
destruction is maximum, as in the case of pyrolyzate production.
Thus, we can use this temperature to define the time when subject
materials are most susceptible to ignition.

Using standard solutions for transient heat conduction in
semi-infinite solids with constant thermal properties, it is
possible to calculate the time at which a material's surface will
attain a specific temperature upon exposure to constant thermal
flux levels. Adjustments should be made to account for
re-radiation heat losses from exposure surfaces and latent heat
processes required to produce pyrolyzates from polymers. With
specific surface temperature, exposure heat flux, and defined
thermal constants, the time required to reach this temperature is
determined by solution of the differential equation for transport
heat flow in a semi-infinite solid:

$$t = \left(\frac{\pi T_s}{2\dot{q}_t}\right)^2 k\rho c_p \tag{1}$$

where

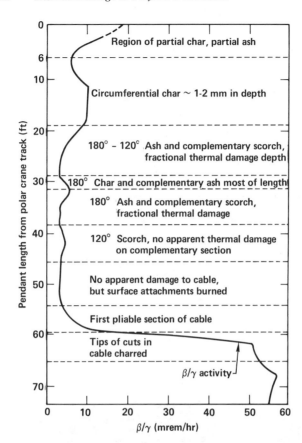

Figure 16. Thermal damage pattern and β/γ activity along Du Pont Hypalon and ethylene propylene covered polar crane pendant.

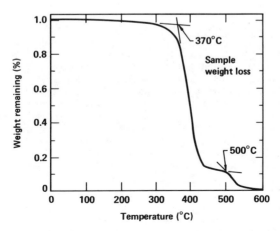

Figure 17. Thermogram of NBS-ABS. In air, 20°C per min heating rate.

$\dot{q}_t$ = total thermal exposure,

T_s = surface temperature,

$k\rho c_p$ = material thermal constants.

Times calculated using this equation should be short relative to those for real materials which experience both thermal and mass convection heat losses. To account for these losses, we adjust q_t by subtracting from it the surface radiation energy at the specified critical surface temperature and the mass convection losses (the product of surface mass loss and latent heat of pyrolysis). The resultant effective energy exposure rate q_e replaces $\dot{q}_t$ in Equation 1, giving a longer time to attain the critical temperature level. Values for time obtained by using both $\dot{q}_t$ and $\dot{q}_e$ in Equation 1 bound the time range between exposure of an inert solid and a solid experiencing both re-radiation and latent heat losses. Critical temperature for the three materials is estimated to be 600°K, and thermal exposure energy is the high value calculated from both convective and radiative exposure during combustion of 8% hydrogen in air ($\dot{q}_t$ = 4.5 W/cm^2). These materials and times to critical weight-loss are

Material	$t_e(\dot{q}_t)$	$t_c(\dot{q}_e)$
Pine wood	5.3 s	9.4 s
PVC	32.0 s	54.7 s
Acrylic	40.0 s	68.0 s

Times to attain critical temperature conditions in these materials are of the same order of duration as those recorded during the hydrogen burn in free volumes of the reactor building. Thus, all susceptible materials exposed to this energy should (and did) experience thermal degradation and/or flaming ignition.

Hydrogen-Fire-Exposure Tests

Thermal constants of most polymeric materials are defined only for virgin compounds. It is virtually impossible to calculate thermal response properties of commercially available polymers because additives, retardants, and fillers modify fundamental properties; however, simple hydrogen-fire-exposure tests may give an indication of accident exposure conditions. To assess this possibility, we conducted selected exposure tests on our exemplar materials using a Meeker burner adjusted to a fully pre-mixed burning mode.(11) Flow was adjusted to produce a measured flame temperature of 833°K (note: during measurement, the 20-mil thermocouple was incandescent, so measured temperature was substantially lower than actual flame temperature). A simple-copper-slug calorimeter measurement of total thermal flux indicated an exposure flux of 6 W/cm^2. This level of flame temperature and thermal flux was within reasonable limits of projected TMI-2 accident measurements and estimated reactor exposure conditions. Thus, resulting data trends should be similar to thermal response variations of materials that suffered hydrogen-flame exposure in the TMI-2 reactor building.

Table I lists results of small-scale hydrogen-exposure tests. Note the time to significant thermal damage is well within times to critical exposure calculated herein. Similarity of thermal damage sustained by materials from the reactor building and those used in the small-scale test were encouraging. Both duration and intensity of test thermal exposure is in the range of estimated thermal fluxes extant during the reactor building burn. Note that these are very simplistic tests. No attempt was made to refine temperature or thermal energy measurement. We have no illusion as to the distribution of convective or radiative contribution from the test burner; however, the results give data trends which are intuitively acceptable.

Conclusions

On the basis of

o   photographic and video surveys of the TMI-2 reactor building interior,

o   visual and photographic analysis of materials extracted from the reactor building,

o   macro- and micro-experiments with materials of composition generically similar to that of extracted TMI samples, and

o   calculations using proposed physical conditons and assumed material properties,

the following conclusions are posed:

1. Hydrogen concentration in the reactor building prior to burn is confirmed to be about 8%, as calculated by analyzers of TMI-2 pressure and temperature records.

2. No preferred path for hydrogen flame propagation has been established, and there is no evidence to preclude hydrogen deflagration throughout the entire free volume of the reactor building.

3. The most probable ignition site for the hydrogen burn is in the basement volume outside of the D-ring: radial location is not defined.

4. Thermal degradation of most susceptible materials on all levels is consistent with direct flame exposure from hydrogen fire.

5. The directional character of damage to lower pendant length
   suggests potential geometric limitation of the hydrogen-
   fire propagation paths.

6. The total burn pattern of the plywood tack board for the
   south-wall telephone on the 347-ft level indicates flame
   propagation through the seismic gap.

7. Lack of thermal degradation to some thermally susceptible
   materials at the 305-ft and 347-ft levels may result from
   preferential moisture absorption, relative to thermally
   degraded materials at adjacent locations. Because of the
   random nature of this evidence, it is not likely that lack
   of damage resulted from selective shadowing.

8. Burn patterns in the reactor building indicate that the
   dome region above the 406-ft level was uniformly exposed to
   direct hydrogen combustion and high heat flux for the
   longest duration. The region between the 406-ft level and
   the top of the D-ring was exposed to directional heat flux
   (most likely from the south and west quadrants); and, the
   damage on the 305-ft level was geometrically similar to
   that above the 347-ft level but less severe.

Table I.   Results of Hydrogen-Fire-Exposure Tests
on Exemplar Materials

Sample	Test	Time (s)	Energy exposure (J/cm^2)	Results of exposure
Polypropy-lene rope	1	12	72	Melted at ends, waxy
	1A	30	180	More melting at ends than test 1, some blending of materials
	1B	27	162	Melting at point of contact, breakage occurred at 27 s into test with moderate pulling force applied
	1C	33	198	More melting than test 1B, break-age occurred at 33 s into test with very little force applied
Telephone receiver cord	2	12	72	Melting, fusing of jacket, conductors exposed, bubbling of clear plastic plug
	2A	30	180	More melting of jacket than test 2, char formation, signs of drip-ping, conductors exposed and ignited at 29 s into test

Table I. Results of Hydrogen-Fire-Exposure Tests
on Exemplar Materials (continued)

Sample	Test	Time (s)	Energy exposure (J/cm^2)	Results of exposure
Telephone dial	3	12	72	Melting at edges, some bubbling
	3A	20	120	Melting at edges, incipient bubbling
Telephone dial (on screen)	4	30	180	Material placed on screen to prevent dripping onto burner): Melted into screen, bubbling
	4A	20	unknown	Inadvertent flame temp decrease approx 30-40°C: Bubbling
	4B	35	unknown	Inadvertent flame temp decrease approx 30-40°C: More bubbling than 4A
Telephone extension line	5	12	72	Melting, charring along edge of cable, bubbling and deformation of clear plastic plug
	5A	20	120	More charring and melting than in test 5; ignited approx 18 s into test
Plywood	6	12	71	Some charring along edges
	6A	20	120	More charring than in test 6, minimal buring through top lamina
	6B	30	180	More charring of top surfaces, outer edges, and corners; splitting of top layer
	6C	60	360	Extreme charring of top surfaces and sides; ashy appearance at corners
Plywood (wet)	7	12	72	No noticeable change
	7A	30	180	Slight char along one edge
	7B	60	360	Charring approx like test 6B
ABS (white material)	8	12	72	Loss of strength; bubbling, slight char, deformation

Table I. Results of Hydrogen-Fire-Exposure Tests
on Exemplar Materials (continued)

Sample	Test	Time (s)	Energy exposure (J/cm^2)	Results of exposure
	8A	20	120	More bubbling, deformation, blackening of approx 74% of surface area
ABS (on screen)	8B	30	180	More bubbling, melted edges, melted into screen, brownish color over surface
	8C	40	240	"
Duct tape	9	12	72	Widespread bubbling, penetration through top (silver) layer
	9A	20	120	More bubbling than in test 9, penetration through top layer
	9B	30	180	More bubbling, charring, melting of adhesive; penetration through top layer
Plywood covered with PE	12	12	72	Plywood covered with single layer of polyethylene one side only: PE burned completely away, charring on 2 opposite edges
	12A	12	72	Plywood covered with double layer of PE on one side only: 25% of PE lost due to drippage and shrinkage; charring along edges of plywood
"	12B	20	120	Double layer of PE on one side of plywood: PE burned completely away; charring at edges and corners of plywood; PE ignited at 15 s into test and one edge of the plywood ignited also
"	12C	12.5	75	Wood placed in PE bag: Bag burned away at approx 7 s; noticeable color change in wood at approx 12.5 s
"	12D	9.5	57	Plywood placed in PE bag: Bag burned away approx 6 s into test; noticeable color change in plywood at approx 9.5 s

Table I. Results of Hydrogen-Fire-Exposure Tests
on Exemplar Materials (continued)

Sample	Test	Time (s)	Energy exposure (J/cm^2)	Results of exposure
"	12E	13	78	Plywood placed in PE bag: Bag burned away approx 6 s into test; noticeable color change in plywood at approx 13 s (this plywood was a darker piece than used in test 12D)
Telephone body	10	12	71	Loss of strength, some wrinkling
	10A	20	120	Leathered appearance, bubbling
	10B	30	180	More bubbling; otherwise same as test 10A
Hose	11	12	72	No noticeable change
	11A	20	120	No noticeable change
	11B	30	180	Some discoloration
	11C	60	360	Charring, slight deformation, melting of outer covering

Acknowledgments

This work was performed under the auspices of the U.S. Department of Energy by Lawrence Livermore National Laboratory under contract No. W-7405-ENG-48. Sections of this paper were originally published in GEND-INF-023, Vol. VI, U.S. Nuclear Regulatory Commission, Washington, D.C. (1983) under DOE Contract No. DE-AC07-76ID01570.
Some of the figures were borrowed and the text paraphrased from References 1, 3, and 5, who devoted extensive time to analyze the hydrogen burn at TMI-2 reactor building. I acknowledge them throughout this chapter, and I hope my interpretation of their analyses is correct.

Literature Cited

1. Zalosh, R. G., et al. "Analysis of the Hydrogen Burn in the TMI-2 Containment"; EPRI NP-3975 Project 2168-1, Final Report, April 1985.
2. VanWitbeck, T. L.; Putman, J. "Annotated Sequence of Events, March 28, 1979"; GPU Nuclear, TDR-044 (1982).
3. Henrie, J. O.; Postma, A. K. "Analysis of the Three Mile Island (TMI-2) Hydrogen Burn"; Rockwell International, Rockwell Hanford Operations, Energy Systems Group, Richland, WA, RHO-RE-SA-8 (1982).

4. Hertzberg, M. "Flammability Limits and Pressure Development in H_2-Air Mixtures"; Pittsburgh Research Center, Pittsburgh, PA, PRC Report No. 4305 (1981).
5. "Flame and Detonation Initiation and Propagation in Various Hydrogen-Air Mixtures, With and Without Water Spray"; Rockwell International, Atomics International Division, Energy Systems Group, Canoga Park, CA, Al-73-29.
6. Lowry, W. E.; Bowman, B. R.; Davis B. W. "Final Results of the Hydrogen Igniter Experimental Program"; Lawrence Livermore National Laboratory, Livermore, CA, UCRL-53036; U. S. Nuclear Regulatory Commission, NUREG/CR-2486.
7. Eidem, G. R.; Horan, J. T. "Color Photographs of the Three Mile Island Unit 2 Reactor Containment Building: Vol.1 Entries 1, 2, 4, 6"; U. S. Nuclear Regulator Commission, Washington, DC, GEND 006 (1981).
8. Alvares, N. J.; Beason, D. G.; Eidem, G. R. "Investigation of Hydrogen Burn Damage in the Three Mile Island Unit 2 Reactor Building"; U. S. Nuclear Regulatory Commission, Washington, DC, GEND-INF-023 Vol. 1 (1982).
9. Hertzberg, M.; Johnson, A. L.; Kuchta, J. M.; Furno, A. L. "The Spectral Radiance Growth, Flame Temperatures, and Flammability Behavior of Large-Scale, Spherical, Combustion Waves"; Proc. 16th Symposium (International) on Combustion, The Combustion Institute, Pittsburgh, PA (1976).
10. Trujillo, R.; Cannon, P. "Cable/Connections Task: Report No. 1 Polar Crane Pendant Cable," unpublished report.
11. Lewis B.; von Elbe, G. "Combustion, Flames and Explosions of Gases"; 2nd ed. Academic Press: New York, 1961; p. 490.
12. Zalosh, R. G., personal communication.

RECEIVED September 10, 1985

Core Damage

G. R. Eidam

Bechtel National, Inc., Middletown, PA 17057

As a result of the March 28, 1979 accident at
the Three Mile Island Unit-2 (TMI-2) nuclear
power plant, it became necessary to characterize
and examine the inside of the reactor vessel.
These activities will directly support defueling
operations, and to efficiently defuel the TMI-2
reactor, it has become necessary to understand the
conditions in the reactor vessel (i.e., damage to
the plenum, depth of the loose debris, etc.).
Knowledge of the conditions within the TMI-2
reactorvessel will also support a second need by
helping the nuclear industry to understand severe
core damage initiation, propagation, and termination,
thus supporting the technical basis for existing
regulations and proving safeness of light water
reactor design and operation.

On March 28, 1979, the Unit-2 pressurized water reactor (PWR) at
Three Mile Island underwent an accident that resulted in severe
damage to the reactor's core. As a consequence of the TMI-2
accident, and for the first time in the commercial Nuclear Power
Industry's history, it became necessary to defuel a severely
damaged reactor. Not knowing the full extent of damage to the
TMI-2 reactor, inspections were needed to better understand the
existing conditions within the reactor vessel in support of this
unique defueling effort.
　　The primary purpose of the core damage examination program is
to obtain necessary data from the reactor vessel in a timely
fashion so that various work plans and tools can be developed in
advance of each portion of the reactor disassembly/defueling
effort. Also, as a result of the accident, numerous aspects of
light water reactor safety have been questioned, and the Nuclear
Regulatory Commission (NRC) has embarked on a thorough review of
reactor safety issues, particularly the cause and effects of severe

0097–6156/86/0293–0087$06.00/0
© 1986 American Chemical Society

core damage accidents. The nuclear community has acknowledged the
importance of examining the TMI-2 reactor in order to understand
the nature of the core damage. A broad spectrum of data is needed
to support the successful resolution of the severe accident safety
questions that have arisen since the accident.

The available data from the accident were studied and a
wide-range of estimates of the potential damage to the reactor
developed. Deformation and/or deterioration of the reactor
components could make reactor disassembly by normal means
difficult, if not impossible. To ensure a successful
disassembly/defueling effort, special equipment and techniques had
to be developed to accommodate the possible abnormal conditions
that could be present. Contingency methods would be needed in the
event that implementation of any proposed techniques prove
unsuccessful or inappropriate. All inspections and data
acquisition programs that have been or will be selected for
implementation must be properly integrated into the disassembly and
defueling operations, or the desired and/or required information
must be obtainable without causing unnecessary disruption and/or
delay of the recovery effort.

The importance of these data is reflected by the fact that,
immediately after the TMI-2 accident, four organizations with
interest in both plant recovery and accident data acquisition
entered into a formal agreement to cooperate in these areas. The
four organizations, commonly referred to as the GEND, include GPU
Nuclear (a subsidiary of the General Public Utilities Corporation,
the plant's owner), Electric Power Research Institute (EPRI),
Nuclear Regulatory Commission (NRC), and Department of Energy
(DOE). These organizations are currently active in reactor
recovery and accident research efforts. The areas to which the
individual GEND organizations are committing available resources
have been defined and coordinated to minimize overlap.

Description of Core Prior to the Accident

To better understand what has happened to the TMI-2 reactor, it is
first necessary to understand the conditions before the accident.
Figure 1 shows a pressurized water reactor and identifies the major
components. Table 1 summarizes the core design data. The fuel in
the TMI-2 core is uranium dioxide (UO_2) powder that has been
pressed, sintered, and ground to form cylindrical pellets 9.4mm
(0.370 in.) in diameter and 17.7mm (0.692 in.) in length. The fuel
pellets are stacked inside Zircaloy-4 cladding tubes (3.9mm (153.2
in.) in length, 10.9mm (0.430 in.) O.D., and 9.6mm (0.377 in.)
I.D.), to create a fuel rod. There are 208 fuel rods in each fuel
assembly, yielding an active fuel length of 3.7m (144 in.). The
TMI-2 reactor contained 177 fuel assemblies (Figure 2). The TMI-2
fuel assemblies contain three enrichment levels of uranium 235
($235U$): 1.98%, 2.64%, and 2.96% (by weight). The configuration
in which fuel assemblies are loaded into the core region depends on
the enrichment level of uranium 235 in each. Fuel assemblies are
constructed with cladding covers, which protect the fuel pellets
from coolant corrosion. The temperatures at which the various
materials in the TMI-2 reactor vessel melt are presented in Table 2.

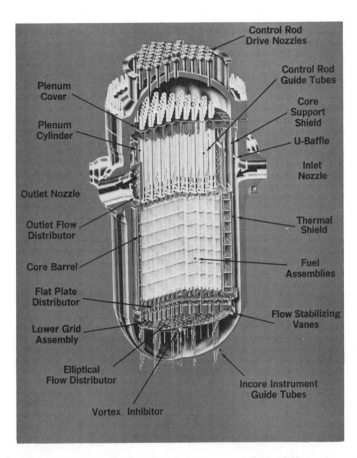

Figure 1. A pressurized water reactor and identifies the major components.

TABLE I.   CORE DESIGN DATA

A. Reactor

    1. Design heat output (MWt)   2772
    2. Vessel coolant inlet temperature, $^{\circ}K$ ($^{\circ}F$)   565 (557)
    3. Vessel coolant outlet temperature, $^{\circ}K.$ ($^{\circ}F$)   592.8 (607.7)
    4. Core coolant outlet temperature, $^{\circ}K$ ($^{\circ}F$)   594.4 (610.6)
    5. Core operating pressure, MPag (psig)   15.1 (2185)

B. Core and Fuel Assemblies

    1. Total fuel assemblies in core   177
    2. Fuel rods per fuel assembly   208
    3. Control rod guide tubes per assembly   16
    4. In-core intrs. positions per fuel assembly   1

C. Total UO_2 Beginning of Life

    1. UO_2 first core (metric tons)   93.1

D. Core Dimensions

    1. Equivalent diameter, m (in.)   3.27
                                   (128.9)
    2. Active height, m (in.)   3.66
                                   (144.0)

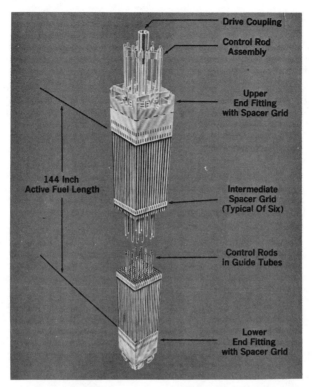

Figure 2. Fuel Assembly - cutaway showing partially inserted
control rod assembly.

TABLE II.   MELTING POINTS OF CORE MATERIALS

Material	Melting Temperature	
	(^{O}K)	(^{O}F)
UO_2	3078	5080
Zircaloy-4 (Fuel Rod Cladding and Guide Tubes)	2123	3362
ZrO_2 (By-Product Metal-Water Reaction)	2988	4919
Inconel 718 (Spacer Grid)	1533–1559	2300–2346
Ag-In-Cd (Control Rod Poison)	1060	1472
304 SS (Cladding of Control Rods and Axial Power Shaping Rods)	1672–1694	2550–2590
SS, grade Cf-3M	1698	2600
Al_2O_3-B_4C (Axial Power Shaping Rods)	2303	3686
UO_2-Gd_2O_3 (2 fuel assemblies contained gadolinia test rods)	3023	4982

First Video Inspection ("Quick Look")

In July 1982, the leadscrew for a control rod drive mechanism
(CRDM) was uncoupled from the control rod spider. The leadscrew
was removed and packaged for further examination and analysis. A
television camera [3.18 cm (1-1/4 in.) in diameter and 35.6 cm (14
in.) long] was lowered through the 3.81 cm (1-1/2 in.) diameter
hole created by the removal of the leadscrew. With the camera at
this level, the spider and fuel assembly end-fittings (Figure 2)
were not seen, indicating that part of the core was gone from its
pre-accident position. The camera was then lowered into the void
in the core, which extended some 1.52 m (5 ft.) below the normal
top space of the core at this location. This inspection afforded
the first visual indication of the severity of the core damage
resulting from the accident. As a result of this "Quick Look"
inspection, the following conclusions were drawn:

o   The upper plenum assembly appeared to be relatively undamaged,
    but some upper end-fittings were stuck to the underside of the
    plenum and partial fuel rods were attached and hanging
    unsupported. The lower ends of several guide tube internals
    had melted.
o   The fuel was severely damaged over a significant portion of
    the core and some of the fuel is in the form of a rubble bed.
o   A large void space exists in the upper center of the core.
o   The rubble bed that exists (directly below) the void space is
    composed of loose material to a depth of at least 35.6 cm (14
    in.). (The depth was determined by mechanical probing with a
    rod).
o   There appeared to be evidence that non-fuel materials
    (stainless steel, Inconel, etc.) had partially melted.

During the initial "Quick Look", a camera was lowered
into three radial locations. First was the center location, in
which a void space was initially discovered. The second location
was halfway out to the periphery, and again the camera was lowered
into a region that extended approximately the same distance into a
void space. The third position was on the core periphery. Here,
the spider assembly was still in place, preventing the camera from
being lowered beyond that point.
     This initial inspection produced approximately 6 hours of
videotape. The videotape allowed project personnel to more
thoroughly examine the extent of core damage revealed during the
inspections.
     The result of this "Quick Look" inspection was the
acknowledgement that the TMI-2 defueling effort would not be normal
and thus normal defueling techniques cannot be used. Also, project
personnel realized that additional characterizations and
inspections would be required to support their
disassembly/defueling effort.

Further Inspections and Examinations

Since July 1982, several types of inspections and examinations
within the TMI-2 reactor vessel have been performed to assist in
defining the current conditions (several years after the
accident). The data obtained are being used to plan for the
ultimate removal of the reactor internals and fuel.
Characterization programs have been performed to gather as much
data and understanding of conditions inside the vessel as early as
possible. Major disassembly (i.e., reactor vessel head removal)
and defueling operations will be based upon the information
provided by these programs. Some of the inspections and
examinations performed to date have included:

o   Removal and analysis of three CRDM leadscrews. Visual
    examinations and chemical and radiological analyses of surface
    deposit were conducted using samples removed from the
    leadscrews. The objectives of the examinations were to
    (a) estimate the maximum temperature experienced along the
    length of the leadscrew in the plenum assembly region, and b)
    determine the extent and nature of core component material and
    radionuclide distributor on plenum assembly surfaces.

o   Axial power shaping rod (APSR) assembly insertion test.
    Following the accident, eight APSRs were in a partially
    withdrawn position (approximately 25% of full travel). A test
    was made to attempt to insert the APSRs to their fully
    inserted position, or at least to a hard-stop position. This
    information was used to assist in determining the physical
    condition of the APSRs and of the damaged reactor core.

o   Uncoupling of all CRDM leadscrews. As a prerequisite to head
    removal, attempts were made to uncouple all CRDM leadscrews.
    The data being sought included: 1) whether the leadscrews were
    present, and (2) whether they could be uncouples. This
    information further aided the task of understanding core
    conditions.

o   Core topography (sonar) scanning. This experiment provided
    the first reliable definition of the shape and size of the
    void region.

o   Reactor vessel head radiological profiling. Gamma radiation
    measurements were taken on the underside of the reactor vessel
    head to determine the potential exposures to personnel during
    the head removal operation. Data were also used to
    radiologically characterize the top of the plenum assembly.
    During this operation, video examinations of the underside of
    the head and the surface of the plenum assembly were performed
    to determine debris accummulation. Also, a debris sample was
    removed for further examination. Several leadscrews were
    parked (raised out of the plenum into the head assembly), and
    the radiation fields on the outside of the service structure
    were monitored to establish the radiological conditions. The
    information produced by this activity was used to help
    determine shielding requirements that would apply when all
    leadscrews were parked for the head removal operation.

o   Rubble bed grab sample removal and analysis. Eleven core
     debris samples were removed from the rubble bed at several
     depths and at two radial locations. Analysis of these samples
     provided data on the extent and nature of damage within the
     core region. The data were used to assist in determining the
     tooling and procedures that will be required to defuel the
     reactor vessel.

o   Detailed camera examinations of the core void region. The
     core void region underwent a video examination to determine the
     amount and condition of material adjacent. In particular the
     information helped recovery engineers to further understand
     the quantity and condition of the material hanging from the
     underside of the plenum, which must be removed prior to plenum
     removal. Also, these data permitted efforts to define the
     type and size of material laying on the surface of the rubble
     bed and of material at the periphery of the void region (these
     materials could fall into the void region if material is
     knocked down from the underside of the plenum). This
     information will be used to support rubble vacuuming and large
     piece removal (pick-and-place) operations.

o   Impact probing of the core rubble bed. The rubble bed was
     probed with a steel rod in 18 radial locations to determine
     the depth and total amount of loose debris. This information
     also assisted in determining what potentially exists below the
     rubble bed.

o   Lower reactor vessel head video inspection. Video inspections
     of the annulus area between the core support assembly (CSA)
     and the reactor vessel lower head were performed to examine
     the amount and condition of debris in the lower head of the
     reactor vessel.

These data acquisition projects, plus others, have provided
the best information to date on the physical conditions inside the
TMI-2 reactor vessel. These data have been to be invaluable to
engineers planning the reactor disassembly/defueling program,
allowing them to design necessary tooling and providing bases for
studies to determine how to safely remove the fuel. The remainder
of this paper focuses on the conditions identified by the above
examinations.

Observed Core Conditions

The TMI-2 reactor vessel internals consist of two major components:
the core support assembly (CSA) and the plenum assembly
(Figure 1). The core was severely damaged by the high fuel
temperatures during the accident. The plenum assembly is located
above the core and was subjected to damage because hot fluids
exiting the core had to pass through the lower portion of the
plenum before flowing to the reactor vessel exit.

Underwater video camera inspections have been the principal
method of determining both core and plenum assembly damage. Based
on these inspections, no significant plenum structure damage is
evident. While no major damage was observed, several examples of
highly localized and variable damage were noticed on the bottom of

the plenum assembly where the tops of the fuel assemblies engage the grid. Also at this level, a number of partial fuel assemblies and end-fittings were found to be attached to the underside of the plenum. Figures 3-6 show some of this damage and some of the material hanging from the underside of the plenum. These figures, along with a number of the remaining figures in this paper, are photographs taken from a video monitor. The only visual information available from inside the reactor vessel (except for samples removed) is on videotape. Several photo-montages have been constructed from the videotape to give engineering personnel a better overview perspective of existing conditions. Thermal expansion buckling was observed in a large fraction of the tubes which normally guide the control rods.

The APSR insertion tests provided the first indication of the condition of the upper reactor internals. Because they performed no safety or criticality function, the APSRs were not inserted during reactor shutdown and remained partially withdrawn (25%) until the APSR insertion tests were performed. The results of the test showed that two of the APSRs could be fully inserted, two could be inserted to the approximately 5% withdrawn position, one could be inserted to the approximately 18% withdrawn position, and, for all practical purposes, the other three could not be inserted at all. The APSRs did move to some degree, indicating that the leadscrews were intact and that the upper plenum guide tubes were not severely damaged. During the process of uncoupling the CRDM leadscrews, the withdrawn APSRs were inserted to approximately their full insertion position. These APSRs were visible in the void region during subsequent examinations.

As a result of core topography (sonar) scanning, it was determined that the void space volume is approximately 9.3 m^3 (330 cu ft.) and the deepest point of the void space is 2 m (6.5 ft) below the bottom of the plenum assembly. The scanning data, coupled with video examinations, indicated that there are no more than 42 standing fuel assemblies. Of these, only 2 assemblies have the potential of being full and intact, 17 have greater than 50% of their horizontal cross section, and 23 have less than 50% of their cross section. Because of the severity of the damage, it is assumed that the two potentially intact assemblies contain ruptured rods which have released noble gases, and that they have incurred other damage and distortion.

The scanning confirmed that large quantities of partial fuel assemblies and end-fittings were suspended from the underside of the plenum assembly. To defuel the reactor, the plenum assembly must be removed from the reactor vessel. Before this can be accomplished, the suspended material from the outside of the plenum must be removed. In November and December 1984, this was done; all suspended material was knocked from the underside of the plenum into the void region. Also, several spiders and end-fittings were found to be on top of the rubble bed (Figure 8) at several locations. The observed hanging fuel assemblies and standing partial fuel assemblies along the edges of the void space appeared to have jagged edges, indicating (with high probability) that they are brittle. The inner assemblies are assumed to be ductile. Many fuel rod pieces are leaning against the core former plates and

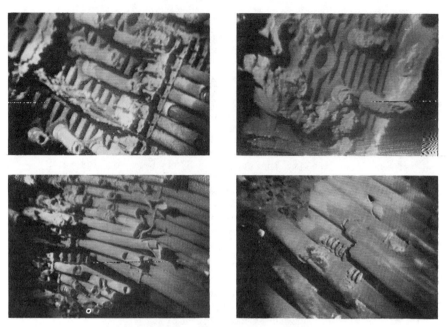

Figure 3 – 6.   Shows some of the damage and some of the material
hanging from the underside of the plenum.

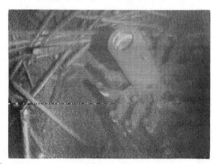

Figure 7.   Relative size of the void space.

laying on the rubble bed. The fuel rod pieces are several
centimeters to a meter or more long. As a result of the efforts to
remove all of the suspended material from the underside of the
plenum, the rubble bed has, been covered with partial fuel
assemblies, spiders, and end-fittings.

Figure 8 shows a pictorial representation of the relative size
of the void space. In several locations, the void space extends to
the plates that compose the core boundary (core-former plates),
which are also visible. It is possible that in several locations
the core former plates are bowed outward slightly, because there
appears to be a larger than normal gap between the remains of
several fuel bundles and the core-former plates (Figure 9). There
are also several "mini bundles" (Figure 10) of fuel assemblies with
fewer than the full number of rods. They are held together with
what appears to be the remains of incore spacer grids that melted
and/or became plugged with small debris. Figures 11-14 are
photographs of some of the items described above.

Up to 50% of the total core inventory of fuel and cladding may
be in the form of loose debris (Figure 14) in the rubble bed. To
better understand the rubble bed, 11 samples were taken at two
radial locations (the center and halfway out to the periphery) and
at various depths. The sampling device penetrated the center
location of the bed to a depth of 77.5 cm (30.5 in.) before
encountering a "hard stop". The rubble bed at the other location
was penetrated 94 cm (37 in.) before the sampler was stopped by
resistance. A sample was taken at each these depths, and all 11
samples were sent off site for (ongoing) analysis.

A summary of the photovisual information obtained from the
first six rubble bed samples is presented in Table 3. Figure 15 is
a summary schematic of the rubble bed grab sample acquisition
project (core location and photos of the first 6 bulk samples).
Small pieces of the sample were identified as fuel pellet
fragments, oxidized cladding, or combinations containing reaction
product material. Many of the particles were composed of ceramic,
resolidified metallic, and "rock-like" materials. The external
surfaces of many particles exhibited open porosity and appeared to
be high-temperature reaction products of two or more core
components. Most of the particles were larger than 1 mm, with less
than 2 weight percent being smaller than 300 m. The following
gamma-emitting radionuclides were present: 60 Co, 106 Ru, 110m Ag,
125 Sb, 134 Cs, 137 Cs, 144 Ce, 154 Eu, 155 Eu, and 241 Am.

A series of pyrophoricity tests were conducted on some rubble
bed samples, but no pyrophoric (combustible) materials were found.
A number of particles have been examined further. One is partially
oxidized cladding with fuel attached. In this particle (Figure
16), molten Zircaloy-4 apparently flowed down the inside surface of
the cladding (in the fuel/cladding gap) after oxidation of the
cladding surface had occured. Dissolution of UO_2 (liquified
fuel) by the molten Zircaloy-4 and oxidation of the previously
molten U-Zr-O mixture is evident around the flow channels. Another
partially oxidized cladding particle contained UO_2 on the
inside, ZrO_2 on the outside, and molten material containing UO_2
attached to the ZrO_2. This material was indicated to be a molten
ceramic that contained about 80 weight percent U and 20 weight

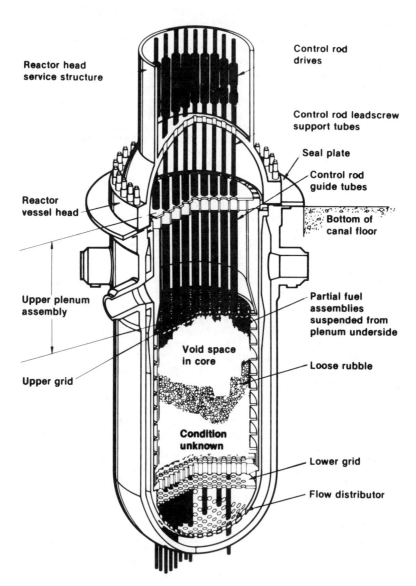

Reactor head
service structure

Control rod
drives

Control rod leadscrew
support tubes

Seal plate

Control rod
guide tubes

Reactor
vessel head

Bottom of
canal floor

Upper plenum
assembly

Partial fuel
assemblies
suspended from
plenum underside

Void space
in core

Loose rubble

Upper grid

Condition
unknown

Lower grid

Flow distributor

Figure 8. On top of the rubble bed, several spiders and
end-fittings were found.

Figure 9. Shows a larger than normal gap between the remains of several fuel bundles and the core-former plate.

Figure 10. Shows several "mini bundles".

Figure 11 - 14. Photographs showing some of the items described in Figure 7 - 10.

TABLE III.   SUMMARY OF PHOTO-VISUAL, ANA GROSS RADIATION LEVELS

Sample Number	TMI-2 Core Location	Location of Sample in Rubble Bed	Gamma Radiation Level at 2.54cm (1") (Rads)	Visual Characteristics
1	H8	Surface	16	A pile of very black, damp debris with a fairly wide range of particle sizes (dimensions ranging from .16 to 1.27cm (1/16 to 1/2 in.); several rounded surfaces; sporadic rust color throughout.
2	H8	7.6cm (3 in.) into debris bed	36	Most particles have brown coloration with wide range of particle sizes (dimensions ranging from 1cm to pan fines) several curved surface; sporadic rust color throughout; several contain substantial surface porosity.
3	H8	56cm (22 in.) into debris bed	36	Very black debris, slightly damp, wide range of particle sizes dimensions .16 to .64cm (1/16 to 1/4 in.), small chunks to fine debris; similar to Sample 1.
4	E9	Surface	3	Thirteen major chunks, dry, black with rust colored sides, basically sharp edges with one or two chunks having rounded edges; dimensions ranging from .64 to 1cm (1/4 to 3/8 in.).
5	E9	7.6cm (3 in.) into debris bed	18	Similar to Sample 4 with the following distinctions: many more pieces; greater size range, .16 to 1cm (1/16 to 3/8 in.); some surfaces more reflective. Again, very dry.
6	E9	56cm (22 in.)	36	Very black debris, small chunks to fine debris, slightly damp, some pieces blackish gray. A couple of pieces resembled metal shards similar to Sample 3.

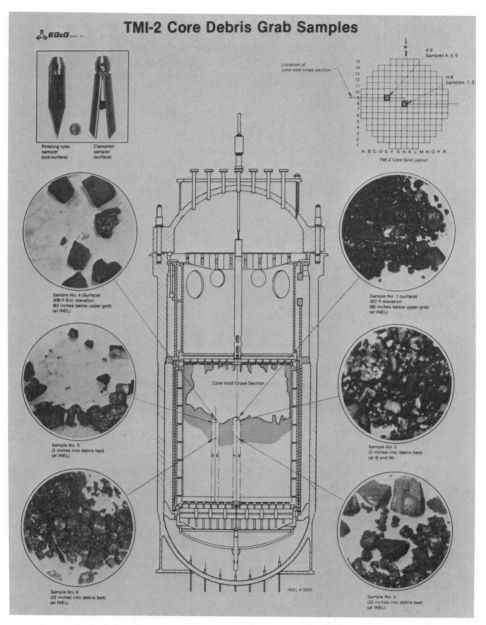

Figure 15. A summary schematic of the rubble bed grab sample
acquisition project.

percent Zr (plus oxygen). The molten material looked like ceramic
and was a single-phase material (no other phases appeared after
etching with the standard Zr and UO_2 etchings). This ceramic
melt particle evidently was exposed to temperatures greater than
2900^oK (4760^oF) and apparently moved to this location on the
outside of an oxidized cladding section and then solidified.
   Analyses of the rubble bed samples have provided valuable data
that supports the defueling effort. The information collected has
assisted engineers in designing defueling tools and a defueling
water cleanup system.
   To further understand the characteristics of the rubble bed, a
pointed rod was used to probe it to determine the depth at 18
radial locations. The tool was lowered until it touched the rubble
bed, and then was inserted into the debris (first manually, and
then forced in with a hammer). The probe tool hit a "hard stop" at
2.54 m $\pm$ .25 m (100 in. $\pm$ 10"in.) below the plenum assembly; this
equates to an approximate rubble bed depth of .9 m (36 in.). Thus,
the approximate dimensions of the rubble bed have been defined and
the presence of hard material beneath the rubble has been
established.
   On February 20 through 22, 1985, a video inspection was
performed in the annulus between the outer surface of the CSA
(Figure 1) and the reactor vessel walls and in the lower head
region. Two radial positions, approximately 180° apart, were
inspected. The annulus between the CSA and the vessel was clean
and free of any large debris. A small amount of small-sized
particulate debris was seen on various horizontal surfaces. The
vent valve, surveillance specimen holder, thermal shield outer
surface, thermal shield support lugs, and seismic strength lugs
were all in normal condition. Bolted joints along the CSA showed
no evidence of any separation. The bottom of the CSA support lugs
and the seismic lugs were still aligned, which indicated that there
was no sagging or slumping of the entire CSA. In the lower reactor
vessel head region, a large quantity of debris was seen. A portion
of the eliptical flow distributor plate (Figure 1) at the bottom of
the CSA was inspected. No damage to this plate nor to a number of
incore instrument nozzles and instrument guide tubes was visible.
   The disposition and composition of the debris was not
uniform. Between 10 and 20 tons of debris were estimated to lie in
the lower reactor vessel head region. The debris has accumulated
to a depth of 75 cm to 100 cm (30 in. to 40 in.) above the bottom
invert of the lower head. The debris pile appeared to be higher at
its periphery than towards its center. The debris appeared to be
segregated radially, with the looser, finer material towards the
center and the bigger material and conglomerations toward the
edges. Figure 17 is representative of the loose debris seen in the
lower head. Formerly liquid material, which appears to have
dropped to the curved area of the reactor vessel lower head and
flowed some distance (Figure 18), is visible in at least two
places. The largest measurable lump of debris was estimated to be
15 cm to 20 cm (6 in. to 8 in.) across.
   While inspecting the flow holes in the flow distributor plate,
engineers noted that about half of the flow holes [15.24 cm (6 in.)
in diameter] contain lumps of agglomerated material (Figures 19 and
20) that appeared to have been molten, congealed, and then broken

84M-877,878   500 μm

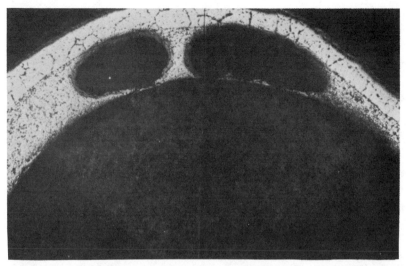

Figure 16. Particle, molten Zircaloy-4 apparently flowed down the inside surface of the cladding (in the fuel/cladding gap) after oxidation of the cladding surface had occured.

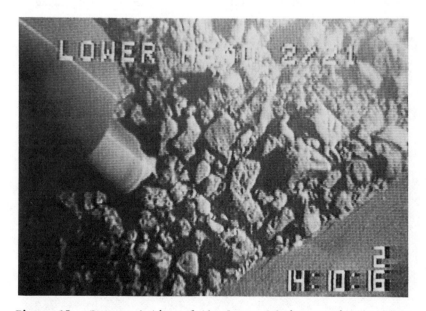

Figure 17. Representative of the loose debris seen in the lower head.

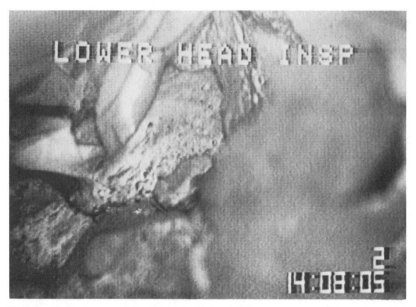

Figure 18. Former liquid material which appears to have dropped
to the curved area of the reactor vessel lower head and flowed
some distance.

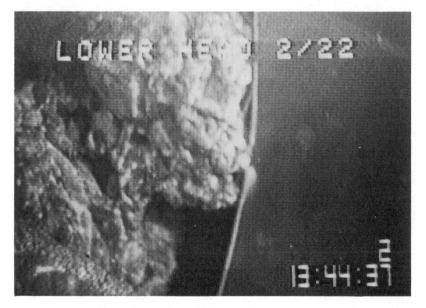

Figure 19. Lumps of agglomerated material that appeared to have
been molten, congealed, and then broken off inside the hole.

off inside the hole. One of these lumps was fractured in several places (Figure 21). As a result of this inspection, additional data from the lower head, particularly in the center of the vessel (which could not be seen in the tests made), will be needed to better understand the conditions that exist. These additional data will be derived from video examinations and samples removed for examination.

Summary

The examinations and characterization activities performed to date have shown that the severity of damage to the TMI-2 reactor core is greater than was anticipated immediately after the accident. It has been determined that little of the reactor disassembly/defueling activities can be performed in a normal manner. Even the removal procedure for the reactor vessel head which used the normal head removal procedure, required modification to take into account unique conditions that exist at TMI-2.

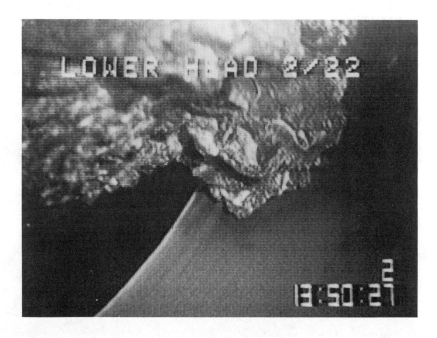

Figure 20. Lumps of agglomerated material that appeared to have been molten, congealed, and then broken off inside the hole.

Characterizations and analyses were needed to support the head
removal operation. Because a vast amount of fission products were
released during the accident, radiological data had to be obtained
to determine whether the reactor vessel head could be removed
without flooding the canal or to establish that the canal would
have to be flooded to protect workers. It was determined that the
head could be removed without flooding the canal, but additional
shielding had to be installed and the workers needed to remain
behind a lead blanket wall, performing most of the operations
semi-remotely with the aid of TV cameras.

Questions still remain about the conditions of the unexplored
portion of the TMI-2 reactor (Figure 7). Additional inspections
and examinations need to be performed to understand what exists
between the rubble bed and the lower head. A wide array of
materials are expected to be encountered, varying from metallic to
ceramic structures. Metallic components may exhibit appreciable
ductility, and other materials may be very hard. Some of the
material may not be fully oxidized. Total comprehension of the
severe damage to the TMI-2 reactor will not be achievable until
the fuel has been removed from the reactor vessel and the reactor
coolant system.

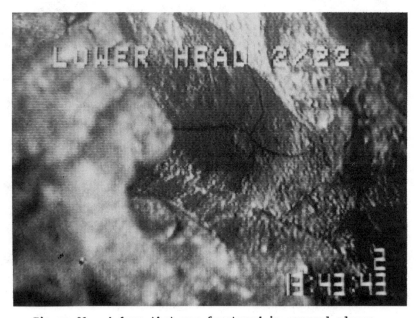

Figure 21. A lump that was fractured in several places.

RECEIVED June 27, 1985

THE CHEMISTRY

6

Water Chemistry

K. J. Hofstetter[1] and V. F. Baston[2]

[1] GPU Nuclear Corporation, Middletown, PA 17057
[2] Physical Sciences Inc., Sun Valley, ID 83353

Prior to the accident, the coolants in the primary and
secondary systems were within normal chemistry specifi-
cations for an operating pressurized water reactor with
once-through steam generators. During and immediately
after the accident, additional boric acid and sodium
hydroxide were added to the primary coolant for control
of criticality and radioiodine solubility. A primary to
secondary leak developed contaminating the water in one
steam generator. For about 5 years after the accident,
the primary coolant was maintained at 3800 ± 100 ppm
boron and 1000 ± 100 ppm sodium concentrations. Dissolved oxygen was maintained <100 ppb by adding hydrazine. Trace quantities of chloride ion (≅1 ppm) and
sulfate ion (≅7 ppm) were found in the coolant. Prior
to decontamination of the coolant, studies indicated
that by maintaining the pH >7.5, corrosion caused by
increased dissolved oxygen levels (up to 8 ppm) and
higher chloride ion content (up to 5 ppm) is minimized.
Chemical control of dissolved oxygen was discontinued
and the coolant was processed. Prior to removal of the
reactor vessel head, the boron concentration in the
coolant was increased to ≅ 5000 ppm to support future
defueling operations. Decontamination of the accident
generated water is described in terms of contaminated
water management. In addition, the decontamination and
chemical lay-up conditions for the secondary system are
presented along with an overview of chemical management
at TMI-2.

The accident at Three Mile Island Unit-2 (TMI-2) resulted in the
release of large quantities of fission products to various reactor
systems and components. These fission products contaminated liquids
in many tanks which in turn produced flooding in the reactor containment building and associated auxiliary buildings. As a result,
water decontamination (i.e., the removal of radionuclides) has been
a major effort in the recovery of TMI-2 (1-5). Because of the large
quantity of "accident generated" water which still contains tritium

(present inventory 1.9 million gallons), the limited post-accident
water storage capacity (2.2 million gallons), and existing regulatory agreements permitting no discharge, sound management of contaminated water required the re-use of previously decontaminated water.
In many cases, chemical additions were required to the post-decon-
tamination water prior to its use in various systems. This paper
discusses the chemical properties of waters both pre- and post-de-
contamination, the chemical adjustments required prior to re-use and
the status of systems. It also discusses the evaluation methods for
the use of chemicals in the decontamination and defueling activities.

History

The reactor coolant received the bulk of the fission products
released from the fuel. In turn, leaking coolant contaminated other
systems. All portions of the coolant purification system accumulated high concentrations of fission products. A coolant leak contaminated the secondary (normally clean) side of one steam generator.
Leakage of coolant from the reactor system resulted in ≅ 650,000
gallons of highly contaminated water collecting in the basement of
the reactor building. Transfers from the reactor building basement
also contaminated large areas of auxiliary buildings, as tanks and
sumps overflowed. One month after the accident, there existed ≅ 1
million gallons of contaminated water containing about one-half of
the core inventory of ^{137}Cs (≅ 360,000 Ci).

While numerous solutions with widely differing chemical compositions
were encountered during the decontamination of this accident generated water, some commonality was evident; namely, the major
solutes were boric acid and sodium hydroxide in varying proportions.
The bulk properties of the solutions were determined and correlated
by a few simple analyses, i.e., pH, conductivity, boron and sodium
concentrations. These chemical analyses were performed, as a
minimum, on all influent and effluent samples taken during liquid
decontamination. Specialized analyses for ionic species present in
trace amounts (e.g., chloride, sulfate, etc.) were also performed on
selected samples. Radionuclide determinations were made by a
variety of radiochemical methods including gamma-ray spectroscopy,
radiochemical separations, liquid scintillation spectrometry and
proportional counting. These analyses and the laboratory facilities
at TMI-2 required to support decontamination activities are discussed in reference 1.

The following discussion is organized in chronological order of
major liquid cleanup operations. Decontamination of general areas
is included in the discussion of each area. Chemical adjustments
required for defueling are presented as they apply to the individual
systems.

Auxiliary and Fuel Handling Buildings

As a result of actions taken to recover from the loss-of-coolant
accident, the Reactor Coolant Bleed holdup Tanks (RCBT) were nearly
filled with contaminated coolant. Other tanks in the auxiliary

buildings were also near capacity. In order to reduce the contaminated water inventory and also gain access to lower elevations in
the auxiliary buildings, a commercial vendor (EPICOR, Inc.) was
contracted shortly after the accident to decontaminate this water.
A water decontamination system was quickly installed in an existing
building and consisted of three ion exchange processing vessels,
operated in series, interposed between two staging tanks. Using the
vendor's proprietary resin mixes, approximately 550,000 gallons of
water from various tanks and sumps in the auxiliary buildings were
"processed" during the period from November 1979 to August 1980.

A summary of the chemical composition of the water processed and the
concentrations of the major long-lived radionuclides is given in
Table I. Most cations and some anions were removed from the liquids
during decontamination. The chemical and radiochemical properties
of the effluent solutions are also summarized in Table I and illustrate the high decontamination factor achieved for the demineralization process.

TABLE I.   A Summary of Chemical Properties of Water in Auxiliary
           Buildings Pre- and Post- Decontamination Using
           EPICOR,Inc. Resins

Analysis	Before	After
pH	8	6
B (ppm)	1000	860
Na (ppm)	300	20
Cl (ppm)	15	<1
SO_4 (ppm)	150	<10
^{137}Cs (μCi/mL)	17	1 E-5
^{90}Sr (μCi/mL)	1	1 E-5

Approximately 290,000 gallons of processed water were transferred to
the Borated Water Storage Tank (BWST), a holding tank reserved for
emergency coolant injection. Approximately 240,000 gallons of the
processed water were transferred to the spent fuel pool while the
balance of the processed water was transferred to in-plant tankage
for interim storage. During these water decontamination campaigns,
nearly 36,000 Ci of ^{137}Cs and 2000 Ci of ^{90}Sr were immobilized on
ion exchange resins. By August 1980, the decontamination and
removal of water in the auxiliary buildings was complete -- permitting less restrictive personnel access to lower elevations in the
buildings and creating necessary freeboard tankage in the reactor
auxiliary systems. Most of the lower elevations were still highly
contaminated due to prolonged exposure to the contaminated water;
however, the general radiation levels were substantially reduced.
The decontamination of these areas is discussed in the next section.

Reactor Building

A specially designed liquid decontamination system, the Submerged
Demineralizer System (SDS), was installed in the spent fuel pool to
process the highly radioactive water in the basement of the reactor
building. By the time the SDS had been installed, the contaminated
water in the reactor building had reached a depth of 8.5 ft. While
the majority of the water in the reactor building had accumulated
during the accident period, the level had continued to rise. The
primary sources of water were leakage from the primary system (69%),
river water in-leakage from the building air coolers (28%), and the
building spray system activated during the accident (3%) (6). The
first sample of TMI-2 reactor building basement water was taken in
August 1979 with analyses reported in reference 7. Prior to SDS
startup, several additional samples were obtained at various locations. Chemical analyses of these samples confirmed continued
leakage from the RCS. A summary of the analyses of the liquid
portions of these samples is given in Table II (6).

TABLE II. Average Analyses Results of Water in Reactor Building
           Basement Before and After Processing Through
                        SDS and EPICOR II

Analysis	Before SDS	After SDS	After EPICOR II
pH	8.5	8.5	4.5
B (ppm)	2100	2100	2000
Na (ppm)	1000	1000	<1
Cl (ppm)	12	12	<1
SO_4 (ppm)	35	35	<10
^{137}Cs ($\mu Ci/mL$)	120	5E-4	1E-5
^{90}Sr ($\mu Ci/mL$)	5	1E-3	1E-5

Processing of this water began in September 1981 and was completed
in March 1982. Decontamination was accomplished using zeolite media
which left the bulk chemical properties (pH, B, and Na) of the solutions unchanged. The operation of the SDS has been described previously (4). The SDS effluent was then further decontaminated (a
process referred to as "polishing") with ion exchange resins in the
original processing system, currently referred to as EPICOR II. The
EPICOR resin mixes were tailored to each batch of SDS effluent such
that the polished effluent could be stored in two 500,000 gallon
storage tanks installed outside the protected area for post-accident
recovery operations. Samples of the SDS effluent were obtained
after every 10,000 gallons of processing and analyzed for chemical
and radionuclide content. The sodium and boron analysis results of
these samples are summarized in Figure 1 for the water pumped out of
the reactor building. Figure 2 shows the chloride, sulfate, and pH
of the water over the same time period.

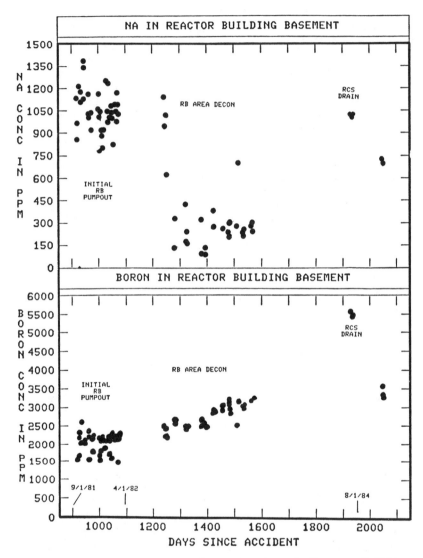

Figure 1. Sodium and Boron Concentrations in Reactor Building Basement Water

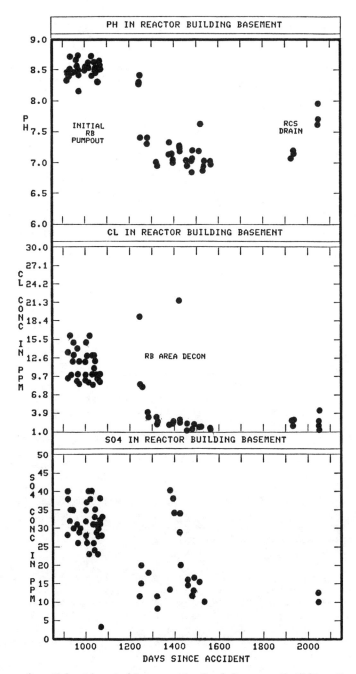

Figure 2.   Chloride, Sulfate, and pH of Reactor Building Basement
Samples

Decontamination of the water in the reactor building basement
immobilized ≅ 280,000 Ci of ^{137}Cs and ≅ 12,000 Ci of ^{90}Sr on zeolite
media. Some of the polished effluent was stored in interim staging
tanks for re-use. After the reactor building was emptied of the
original accident generated water, large scale flushing of the
reactor building surfaces began in July 1982 using the previously
decontaminated reactor building basement water. As the boric acid
was not substantially reduced during liquid decontamination, the
boric acid concentration of the water used for the flushing operations was ≅ 2000 ppm. As a result of the flushing, the reactor
building basement water level increased and was periodically pumped
through the SDS for decontamination resulting in fluctuating water
levels in the reactor building. The boron concentration in the
water increased as seen in Figure 1. As decontamination of reactor
building surfaces proceeded using high pressure (10,000 psi), high
temperature (140 °F) flushing operations, evaporation losses concentrated the boric acid solution. In contrast, as all sodium was
removed during polishing, the sodium concentration generally decreased in the reactor building water. To further complicate the
picture, part of the reactor coolant was occasionally drained to the
reactor building to support planned activities. This accounts for
the occasional high sodium and boron concentrations as noted in
Figure 1. As a result of the decrease in Na concentration and increase in B content, the pH in the reactor building basement reached
a minimum of ≅ 7.

The sulfate and chloride concentrations decreased during the initial
decontamination operations since both were removed by the EPICOR II
polishing operations. However, as more aggressive flushing techniques were used to remove more of the surface contamination, the
sulfate concentrations have occasionally increased presumably due to
dissolution of concrete fines (see increases in Figure 2).

To date, ≅470,000 gallons of previously decontaminated water have
been used for large scale decontamination of reactor building
internal surfaces and has been subsequently re-processed through
SDS. Approximately 15,600 Ci of ^{137}Cs and 4400 Ci of ^{90}Sr have been
removed by SDS/EPICOR II processing of this water. Completion of
large scale decontamination of the reactor building coupled, with
liquid removal from the basement has accomplished substantial
reductions in general area dose rates and contamination levels.
These milestones have been necessary prerequisites to reactor vessel
head removal and defueling operations.

During this period of time, decontamination of the lower elevations
of the auxiliary buildings was also accomplished using the previously decontaminated water. The flush water collected in sumps and
tanks and was re-processed through the SDS and/or EPICOR II systems.
To date, a total of 75,000 gallons of water has been used for
decontamination of these buildings and has eliminated high contamination levels in accessible areas of their lower elevations.
System internal flushes have also been performed using processed
water. Unrestricted access to most areas of the auxiliary buildings
is now permitted.

Reactor Coolant System

Fresh coolant was injected into the Reactor Coolant System (RCS) during the accident to balance the coolant losses due to leakage. The pH (by addition of sodium hydroxide) and boron content in the coolant were increased to maintain subcriticality and minimize the release of radioiodine. Figure 3 displays the boron and sodium concentrations in weekly coolant samples for one year after the accident. After the initial post-accident chemical adjustment, the average pH of the coolant was maintained at 7.8 for about 5 years. The concentrations of other species measured in the coolant samples prior to coolant processing are summarized in Table III. For about 3 years after the accident dissolved oxygen control was accomplished by addition of hydrazine to the makeup system.

TABLE III. Chemical Properties of Reactor Coolant Prior to Processing and to Vessel Head Removal

Analysis	Prior To Processing	Prior To Head-Lift
pH	7.8	7.6
B (ppm)	3800	5050
Na (ppm)	1000	1500
Cl (ppm)	1.0	1.6
SO_4 (ppm)	8	$\cong 7$
Dissolved-O_2 (ppm)	0.1	$\cong 2$
^{137}Cs ($\mu Ci/mL$)	13.5	0.81
^{90}Sr ($\mu Ci/mL$)	16.0	3.9

In July 1982 decontamination of the coolant began by performing bleed-and-feed operations with SDS processing of the letdown batches. At that time, addition of hydrazine to the feed solutions was discontinued. The combination of high pH and low chloride content was considered sufficient to prevent intergranular corrosion of primary coolant system internals. The makeup and letdown tanks used for the bleed-and-feed operations were maintained under a nitrogen atmosphere to minimize oxygen in-leakage, however. The details of the early RCS processing batches are given in (9).

Radionuclides continue to leach out of the damaged fuel into the coolant (consistent with congruent dissolution data from laboratory studies on Spent Fuel Storage programs - 10). As makeup to compensate for RCS leakage continued at various rates, the accumulation rates of radionuclides in the coolant were estimated for selected periods of time but are difficult to accurately quantify. Figure 4 shows the concentrations of the major radionuclides in the coolant for $\cong 6$ years after the accident. Various coolant decontamination flow paths have been used due to changes in the primary system status. For instance, bleed-and-feed operations have been performed after the RCS was in a coolant-lowered and open-to-the-atmosphere condition.

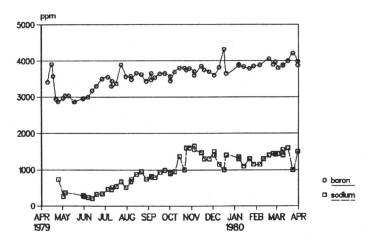

Figure 3.   Sodium and Boron Concentrations in Reactor Coolant for
One Year Following Accident

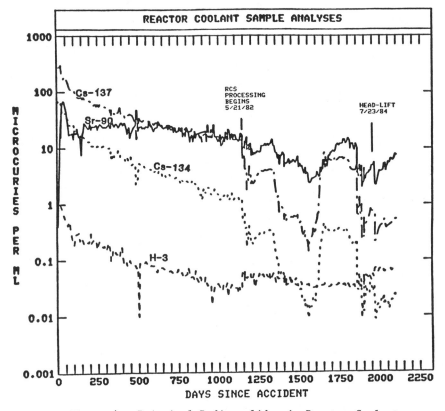

Figure 4.   Principal Radionuclides in Reactor Coolant

As defueling planning proceeded, in order to assure an adequate
margin of subcriticality safety, the decision was made to increase
the boron concentration. The operational target established was
5050 ppm B with a corresponding pH greater than 7.5. Both the RCS
and BWST were adjusted to 5050 ppm B and 1500 ppm Na by adding boric
acid and sodium hydroxide. A total of 17 tons of boric acid was
added. The boron concentration in the coolant was increased gradually by chemical addition to the makeup and by performing batch
bleed-and-feed operations. The chemical properties of the coolant
prior to reactor vessel head removal are given in Table III. The
change in boron concentration in the RCS is plotted in Figure 5.
Because the removal efficiency of ^{137}Cs and ^{90}Sr by zeolites is
strongly dependent on the sodium ion concentration, it was desirable
to add the minimum quantity of sodium hydroxide and yet maintain a
pH greater than 7.5. The pH of the coolant was intentionally
lowered to ≅7.6; this decrease in pH is shown in Figure 5. The
chloride concentration was observed to increase during the chemical
adjustment due to chloride impurities in the chemicals added. The
pH, B, and Na relationship of Mesmer et.al., (11) was used to
calculate the chemical additions which were made from a small (6000
gallon) heated tank. The solubility of sodium hydroxide and boric
acid in water was determined using the solubility isotherms published by U.S. Borax, Inc. (12). Previously decontaminated water
was used for all chemical additions.

Once-Through Steam Generators

Sampling of the secondary side water in one Once-Through Steam
Generator(A-OTSG) occurred frequently after the accident. Higher
than expected conductivity and sodium concentration were observed in
the samples immediately after the accident. Traces of ^{137}Cs were
also found in the water, presumably due to the use of contaminated
feedwater. By early June 1979, the water quality had substantially
improved by using the condensate polishing system. Hydrazine and
ammonia were injected in January 1981 to achieve standby conditions.
The changes in pH, cation conductivity and hydrazine for A-OTSG
prior to processing are shown in Figure 6. Also shown are the
acceptable ranges recommended by the manufacturer for standby
conditions.

Sampling of the other steam generator (B-OTSG) was possible only
until September 14, 1979, when a sample valve in the reactor building failed. The results of analyses of the B-OTSG samples showed
conductivities of ≅30 umho/cm with a sodium content of ≅5 ppm and a
pH ≅9.8. During the accident, the activity level of the secondary
side water indicated a leak of primary coolant into the B-OTSG.
Circulation was stopped at this time. No additional samples of the
B-OTSG were obtained until December 1982.

Preparations for defueling required lowering of the water in the
secondary side of both steam generators to prevent criticality
through boron dilution via a secondary to primary leak. Prior to
draindown, it was decided to purify the water in the secondary
system using ion exchange resins and to chemically adjust the water
to minimize corrosion in the partially drained steam generators.

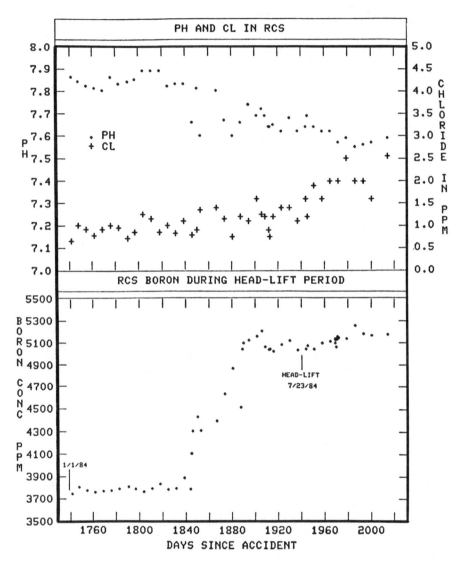

Figure 5. Changes in pH, Chloride and Boron Concentrations in
Reactor Coolant Prerequisite to Reactor Vessel Head Removal

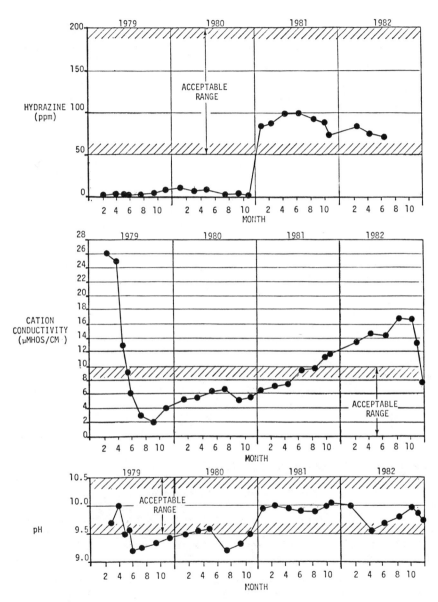

Figure 6. Chemical Analyses of OTSG-A Samples Prior to Decontamination and Chemical Cleaning

A system was designed and installed to fill and recirculate the
OTSG's thereby wetting all internal surfaces. Following recircula-
tion and sampling, the water was circulated in a bleed-and-feed mode
through a demineralizer vessel containing ion exchange resin selected to remove chemical impurities and radionuclides. After cleanup,
hydrazine and ammonia were added with circulation and sampling to
confirm layup conditions. A summary of initial and final conditions
for both steam generators is given in Table IV. It is evident from
these data that significant contamination of the water occurred as a
result of wetting the internal surfaces. The steam generators were
then drained under nitrogen cover gas to a level just below the
anticipated primary coolant level for defueling. They have remained
in this condition since February 1983.

TABLE IV.   Chemical Analyses of Once Through Steam Generator
            Samples Taken Pre- and Post- Decontamination

Analysis	-----A-OTSG---------		-----B-OTSG-----	
	Before	After	Before	After
pH	9.92	9.54	9.85	9.5
COND (μmho/cm)	40	4.3	35	<1
NH_3 (ppm)	2	5	4	5
N_2H_4 (ppm)	140	100	0.1	103
Na (ppm)	17	0.32	5	0.3
Fe (ppm)	0.10	–	<0.10	<0.10
Cr (ppm)	0.005	–	<0.10	<0.10
Ni (ppm)	0.005	–	<0.10	<0.10
Cl (ppm)	0.10	0.23	0.109	0.037
PO_4 (ppm)	0.251	–	0.070	0.140
NO_3 (ppm)	0.012	–	0.127	0.080
SO_4 (ppm)	1.979	–	1.750	0.118
^{137}Cs (μCi/mL)	8.2 E-7	1.3 E-5	8.5 E-3	1.8 E-3

Makeup and Purification Demineralizers

The coolant purification system was in service for only several
hours during the accident. This system consists of two demineralizer vessels each containing $\cong$50 ft^3 of organic ion exchange resin
and several filters. The resins were exposed to highly radioactive
coolant for a only short period ($\cong$2 hours). The system then
remained isolated for $\cong$4 years.

Characterization of the demineralizer resins was accomplished by a
variety of non-destructive assay techniques using remote technology.
Quartz fiber optics were used to visually examine the resins. One
demineralizer was found to be dry and contained crusted resin. The
other was found to contain resin and highly radioactive water (2.64
mCi ^{137}Cs/mL). Samples of the liquid and the resins were obtained
for laboratory analyses. A summary of these results is given in
Table V. The liquid was found to contain a high concentration of

sulfate ion probably due to decomposition of the resin's sulfonate
functional groups. A high level of organics were also found in the
sample presumably from the degradation of the polystyrene resin. It
is postulated that the initial high concentration of sulfate in the
reactor coolant originated from the resin degradation. Early
samples of coolant also contained oxalate ion as a suspected resin
degradation product.

The detailed characterization of the demineralizer resin and the
removal of the ^{137}Cs from the resin by gradient elution are described in detail in a later paper at this conference. As these
resins accumulated radiation dose rates in excess of 2000 megarads
and experienced high temperatures, additional samples will be
removed for careful study at a later date.

TABLE V.   Chemical Analyses of Makeup and Purification
Resins and Liquids

Analysis	-B- Demineralizer Liquid	Resin	-A- Demineralizer Resin
pH	5.3	–	–
B (ppm)	3000	>200	>200
C (ppm)	1000	>10%	>10%
Na (ppm)	7000	>1000	>1000
SO$_4$ (ppm)	9600	15000	10000
U (ppm)	0.064	1620	2420
Pu (ppb)	0.72	3550	4600
^{137}Cs (mCi/mL)	2.64	11.2	4.57
^{90}Sr (mCi/mL)	14	490	–

Controls on Use of Chemicals

Because the decontamination of systems and areas is essentially a
chemical problem, a procedural review and approval system was developed to control the use of chemicals at TMI-2. Prior to use, each
chemical must be formally evaluated against multiple criteria. A
methodology was developed to evaluate the safety of chemicals with
the goal of preventing harm to personnel, property and the environment during chemical transport, use, storage and disposal. Chemicals are evaluated with emphasis on industrial safety, material and
chemical compatibility, corrosion, reactivity, radioactive and
hazardous waste management, and criticality. "Chemical coders"
review the proposed use of chemicals and designate hazard types and
levels from chemical evaluations, vendor data, OSHA Material Safety
Data Sheets and other references. These chemical evaluations become
part of any work instructions where chemicals are to be used. Any
special limits and precautions are annotated. Approval to use the
chemicals is made only after a comprehensive examination of benefits
and risks.

Examples of chemicals that have been approved for use at TMI-2 include: concentrated H_3PO_4 and FREON for the electropolishing decontamination of small tools and components (liquid waste is solidified
prior to disposal); TRITON X-100, a surfactant used in localized
decontamination (limited to 0.1% solution for disposal in liquid
radwaste system); sulfamic acid for the decontamination of concrete
surfaces (localized use with solidification of liquid waste); etc.
Future large scale decontamination chemical use will be evaluated by
these same criteria.

Summary

The chemical conditions of water used to support defueling and
decontamination of TMI-2 will continue to be monitored and changed
as required. Presently, boric acid and sodium hydroxide are being
added to $\approx$350,000 gallons of previously decontaminated water in
order to fill the fuel transfer canal and fuel storage pool in
support of defueling.

The management of contaminated water is a high priority effort in
the recovery of TMI-2. To date, nearly 3.3 million gallons of water
have been processed through SDS and EPICOR II. Of the present 1.9
million gallons of accident generated water inventory, approximately
80% has been processed in a once-through mode. The remaining water
is continually being re-used and re-processed to support decontamination and defueling activities resulting in a considerable cost
savings and minimizing the increase of contaminated water inventory.

Planning is underway to explore chemical reagents for internal
systems decontamination. The use of chemicals will be tightly
controlled to prevent injury to personnel and to prevent damage to
systems. Analysis methods for chemical species will be developed to
support these operations as defueling and decontamination operations
require.

Literature Cited

1. K. J. Hofstetter, C. G. Hitz, K. L. Harner, P. S. Stoner, G.
    Chevalier, H. E. Collins, P. Grahn and W. F. Pitka, "Chemistry
    Support For Submerged Demineralizer System Operation at Three
    Mile Island", Analytical Chemistry in Energy Technology, ed. by
    W. S. Lyon, Ann Arbor Science, p. 301 ff, (1982).

2. K. J. Hofstetter, C. G. Hitz, T. D. Lookabill and S. J.
    Eichfeld, "Submerged Demineralizer System Design, Operation and
    Results", Proceedings of ANS-CNA Joint Topical Meeting on
    Decontamination of Nuclear Facilities, Niagra Falls, Vol. 2, p
    5-81 (1982).

3. K. J. Hofstetter and C. G. Hitz, "Processing of the TMI-2
    Reactor Building Sump and Reactor Coolant System", Proceedings
    of 1982 ANS Winter Meeting, Washington, DC, Vol. 43, p 146
    (1982).

4. K. J. Hofstetter and C. G. Hitz, "The Use of the Submerged
    Demineralizer System at Three Mile Island", Separation Science
    and Technology, 18 (14 & 15), p 1747-1764 (1983).

5. H. F. Sanchez and G. J. Quinn, "Submerged Demineralizer System
    Processing of TMI-2 Accident Waste Water", US Department of
    Energy Report GEND-031 (February 1983).

6. C. V. McIsaac and D. G. Keefer, "TMI-2 Reactor Building Source
    Term Measurements: Surface and Basement Water and Sediment", US
    Department of Energy Report GEND-042 (October 1984).

7. D. O. Campbell, E. D. Collins, L. J. King and J. B. Knauer,
    "Evaluation of the Submerged Demineralizer System (SDS) Flow-
    sheet for Decontamination of High-Activity-Level Water at the
    Three Mile Island Unit Nuclear Power Station", Oak Ridge
    National Laboratory Report ORNL/TM-7448 (July 1980).

8. "Reactor Building Radiological Characterization" Vol. I and II,
    TMI-2 Technical Planning Department/GPU Nuclear Internal Report
    TPO/TMI-125 (Unpublished).

9. K. J. Hofstetter, C. G. Hitz, V. F. Baston, A. P. Malinauskas,
    "Radionuclide Analysis Taken During Primary Coolant Decontam-
    ination at Three Mile Island Indicate General Circulation",
    Nuclear Technology, Vol. 63, No. 3 (December 1983).

10. V. F. Baston and K. J. Hofstetter, "Long Term Appearance Rate
    of Radionuclides in TMI-2 Coolant", Proceedings of 1984 ANS
    Winter Meeting, Washington, DC, Vol. 47, p 111 (1984).

11. R. E. Mesmer, C. F. Bates, Jr., and F. H. Sweeton "Acidity
    Measurements at Elevated Temperatures. VI. Boric Acid Equili-
    bria", Inorganic Chemistry Vol. II, No. 3, p 537 ff (1972).

12. "Solubility Isotherms in the System Borax-Boric Acid-Water at
    0-94 °C", data provided by US Borax (private communication).

13. M. K. Mahaffey, E. J. Renkey, W. W. Jenkins, L. M. Martinson
    and R. D. Hensyel, "Resin and Debris Removal System Conceptual
    Design - Three Mile Island Nuclear Station Unit 2 Makeup and
    Purification Demineralizer", Hanford Engineering Development
    Laboratory Report HEDL-7335 (March 1983).

RECEIVED June 27, 1985

7

Adherent Activity on Internal Surfaces

V. F. Baston[1] and K. J. Hofstetter[2]

[1] Physical Sciences Inc., Sun Valley, ID 83353
[2] GPU Nuclear Corporation, Middletown, PA 17057

The adherent radionuclide cesium activity observed on the
TMI-2 internal surfaces is consistent with the formation
of cesium silicate within the surface duplex oxide on the
metal surfaces. Chemical and physical examinations (ES,
SEM, XRD, AES, metallography, hardness, particle analyses, etc.) of debris and metallographic sections from
internal reactor components removed from the Three Mile
Island Unit-2 have been performed. The principal components removed were three leadscrews (H8, B8, E9), a small
bottom end section of the H8 Leadscrew Support Tube
(LST), and a resistance thermal detector thermowell.
Examination of the leadscrews indicate that radionuclides
and original core materials are associated with Loosely
Adherent surface Deposits (LAD) and Adherent surface
Deposits (AD). The LAD generally comprises material
associated with or evaporated from the Reactor Coolant
System (RCS) fluid whereas the reaction products between
accident material and the leadscrew surface, and in-service corrosion film constitute the AD. The AD contains
the principal gamma activity (up to 3 mCi Cs-137/cm^2)
associated with the leadscrew surfaces. An aggressive
solution (HNO$_3$-HF) was required to remove the cesium
activity from the AD. Hence, surface characterization
analyses were performed to elucidate the chemical nature
of the AD. Results indicate there is a dense region with
a cermet type structure of various elemental compositions
at the LAD/AD interface. The surface analyses also
indicate that silicon and cesium levels are spatially
correlated, a fact pertinent to understanding the adherent cesium activity.

On 28 March 1979 the Three Mile Island Unit 2 reactor under went a
loss of coolant accident that resulted in core damage with subsequent
release of fission products in the reactor vessel (1). In July 1982,
three control rod drive leadscrews were removed from the top of the
reactor vessel (2). The H8 leadscrew is from the center of the
vessel, B8 is near the outer edge, and E9 is approximately mid-radius

0097–6156/86/0293–0124$06.50/0
© 1986 American Chemical Society

(Figure 1). The axial distribution of gamma emitting radionuclides was determined by segmented gamma ray spectroscopy (3). The gamma scans indicated generally similar radionuclide activity profiles along the entire 24 foot length of the leadscrews (Figure 2). The H8 leadscrew was selected for the initial laboratory examinations due to its location at the center of the reactor vessel. Three sections, approximately one foot long, (H8-4, H8-5, H8-6) were cut from the middle threaded portion of the 24 foot H8 leadscrew (stainless steel 17-4PH) for laboratory examination. The sections were covered with a black coating overlayed with heavy rust colored deposits. The three sections ranged in average ^{137}Cs surface activity from 290 to 1100 uCi/cm^2. Brushing of the threaded leadscrew surface removed loose material/debris revealing a "glassy" surface.

As a means of characterizing the chemical nature of the surface material, a series of solution chemistry tests were performed at TMI-2 on one section of the H8 leadscrew (4). An aggressive solution (HNO$_3$-HF) was required to achieve dissolution of the cesium activity (Figure 4). The identification of adherent cesium activity on the leadscrew surface provided the basis for subsequent detailed laboratory examinations on the other H8 sections as well as the radially distant and potentially different B8 leadscrew.

Portions of these leadscrews were subjected to extensive laboratory analysis (4-8). The results provide both a basis for characterization of the decontamination requirements and identification of the physical and chemical conditions during the accident. Chemical and radiochemical analyses also provide a basis for characterizing the extent of radionuclide retention within the reactor vessel and hence its effect on the source term which is defined as the amount and type of radionuclides available for release to the environment. While the majority of existing data for TMI-2 primary system surfaces consists of the H8 and B8 leadscrew data, significant information was derived from two additional components of the Reactor Coolant System (RCS): 1) a section of the H8 Leadscrew Support Tube (LST) from the H8 control rod drive mechanism (9) and 2) the Resistance Temperature Detector (RTD) thermowell (10) from the Once Through Steam Generator - OTSG-A.

The results of the surface examinations of these primary system components are generally consistent while specific results may diverge somewhat due to the examinations being carried out by different organizations utilizing various techniques. Consequently, correlation between the reported analyses involves some interpretation; hence, general trends are more relevant than specific values reported. This paper provides some general comparisons for selected data from the H8 and B8 leadscrews as well as some for the LST and RTD thermowell. The significance of these comparisons is the quantity of adherent cesium activity retained within the reactor vessel and the future decontamination requirements for the reactor vessel internal surfaces. The consistency between adherent cesium results from experimental studies on stainless steel systems (11) and TMI-2 surface characterization studies (8) are noted and presented in the section on Cesium/Stainless Steel Interaction Laboratory Studies.

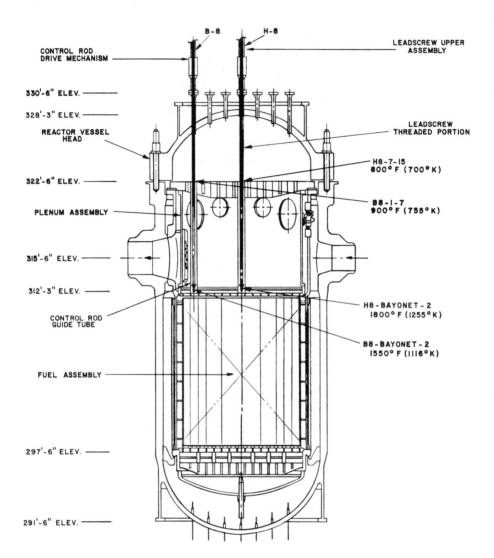

Figure 1   H8 Leadscrew Assembly In Tripped Position With
Leadscrew Sample Temperature Results.

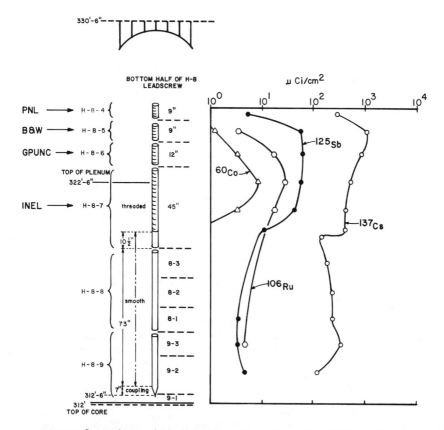

Figure 2   Radionuclide Activity Along Length of H8 Leadscrew.

This paper does not provide a full summary of available analyses or comparisons of the voluminous radionuclide data for samples discussed herein. However, the selected data illustrate the general trends. The data support the conclusions that the core damage occurred under highly heterogeneous conditions based on the disparity in the composition of the subsamples. The detailed results of the laboratory analyses of the TMI-2 components can be found in the principal references 3-10.

Selected Analyses and Data Comparison

Leadscrew Analyses. The reactivity of the core under operating conditions is regulated by control rod assemblies. The leadscrews are part of the drive system used to raise and lower the control rods in the reactor core region. The leadscrews H8, E9, and B8 were the first major components to be removed from the TMI-2 reactor vessel and hence the focus of extensive laboratory examination.

The analytical results are generally presented chronologically as this order of information reflects the basis for both the direction for additional analyses and hypotheses presented. The initial solution chemistry tests performed at TMI-2 provided the unanticipated results of an adherent cesium activity and generally set the initial direction of laboratory examinations to ascertain the nature of the adherent cesium. Our discussion of analyses begins with these tests.

The solution chemistry tests involved submerging one of the three H8 leadscrew sections, H8-6, (the average ^{137}Cs surface activity was 840 uCi/cm^2) in successive solutions arranged in increasing order of chemical aggressiveness. The aqueous media were: 1) demineralized water, 2) borated water (3500 ppm boron as boric acid adjusted to pH 7.5 with NaOH and containing 1% TRITON X-100 surfactant), 3) sodium carbonate and hydrogen peroxide solutions (5.0 Wt.% Na_2CO_3 and 1% H_2O_2), 4) solutions used in a standard two step decontamination process-APC ((i) 10% NaOH - 3% $KMnO_4$ and ii) oxalic acid (25 g/L) - dibasic ammonium citrate (50 g/L), and 5) nitric and hydrofluoric acid (10 Wt.% HNO_3 - 0.1M HF).

Figure 3 shows the results of the successive treatments. The carbonate/peroxide solution dissolves fuel and associated debris as shown by the sharp increase in the fuel related nuclides (i.e. ^{239,240}Pu and ^{238}U). The two step APC process is shown to remove corrosion products and associated radionuclides. The nitric-hydrofluoric acid solution removes base metal and is the only solution which effectively removed the radionuclides on the leadscrew surfaces.

Each solution was analyzed for alpha, beta, and gamma emitting radionuclides in both the liquid and insoluble phases. The soluble portion was also analyzed for elemental content by source excited X-ray fluorescence (XRF). The radionuclides remaining on the leadscrew after each soaking operation were determined by gamma-ray spectrometry (4). From the iron concentrations, determined by XRF, the depth of base metal removed by all solutions was estimated to be approximately 0.3 micrometers. The solution chemistry tests provided the basis for the following tentative conclusions:

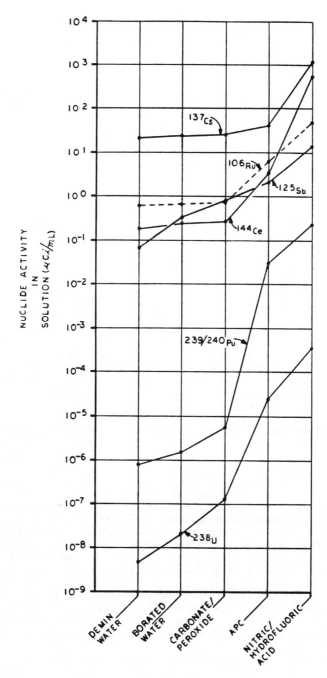

Figure 3   Cumulative Concentration Of Selected Radionuclides Removed By Various Soak Solutions.

130 THE THREE MILE ISLAND ACCIDENT

o $^{137/134}$Cs constitute the principal gamma source on
the leadscrew surface (Table I).

o An aggressive solution was required to effect cesium
removal from the leadscrew.

Table I. Percentage Gamma Contribution for Selected Radionuclides
Associated with the TMI-2 Leadscrews LAD/Brushoff Debris

Isotope	H8-6 (%)	H8-5 (%)	H8-4 (%)	H8-7 (%)	B8-1 (%)
^{144}Ce	1	5	8	27	N.D.
^{137}Cs	78	77	58	40	91
^{134}Cs	10	5	13	2	5
^{125}Sb	1	10	8	13	4
^{106}Ru	1	2	N.R.	14	N.R.
^{60}Co	1	1	12	4	1

N.D. = Not Detected, N.R. = Not Reported

Consequently, detailed laboratory examinations and surface characterizations were conducted on B8 as well as on the remaining H8 leadscrew sections (4-8). Our initial hypothesis on the mechanism for
cesium adherence was that it was associated with a boron film which
led to high resolution SEM studies (8). The hypothesis was based on
the "glassy" surface appearance, chemically adherent cesium, and
known boriding technology. However, more recent laboratory experiments indicate that the adherent cesium is more closely associated
with silicon.

The initial detailed laboratory examination and surface characterization studies (performed on the threaded H8 leadscrew section H8-5)
involved chemical and radiochemical analyses, optical microscopy,
metallographic examinations, scanning electron microscope (SEM)
examinations, microchemical and X-ray diffraction examination, as
well as particle size determinations. These initial microscopic and
chemical examinations indicated that three distinct layers with
varying elemental composition covered the leadscrew base metal
(Figure 4 and Table II). These layers were categorized as: (1)
in-service Metal Oxide (MO) adjacent to the base metal, (2) Adherent
surface Deposits (AD) adjacent to the MO, and (3) Loosely Adherent
surface Deposits (LAD) adjacent to the AD (5).

Examinations were performed on the LAD utilizing emission spectroscopy (ES), atomic absorption (AA) and X-ray fluorescence (XRF) techniques for elemental compositions and nuclear techniques (gamma spectrometry and liquid scintillation) for selected radionuclides (5).

Once the AD layer was removed there was no residual cesium activity.
These analyses indicated that the majority of the gamma activity
associated with the LAD and AD layers was due to the cesium radionu-

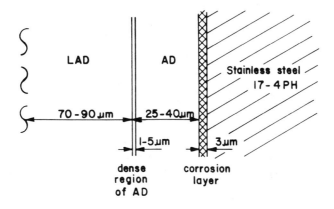

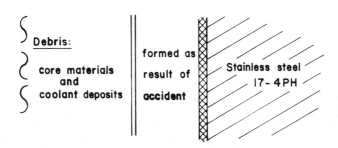

Figure 4   Summary Of Analysis Results Depicting LAD/AD Layers In Schematic Form.

Table II.   Comparison of Analyses Reported for the Loosely Adherent
Surface Deposits (LAD) and Adherent Surface Deposits (AD) from
the TMI-2 Leadscrews

| | ----------LAD---------- | | | | --------AD-------- | | |
| | H8-6 | H8-5 | H8-4 | H8-7 | H8-6 | H8-5 | H8-7 |
Element	Wt.%	Wt.%	Wt.%	Wt.%	Wt.%	Wt.%	Wt.%
Fe	33.8	31.5	9.4	37.0	26.1	23.1	41.6
Zr	25.9	8.1	7.2	0.4	52.6	1.0	0.2
U	22.8	15.0	N.R.	N.D.	9.2	0.9	N.R.
Cr	3.8	3.2	0.8	22.0	2.6	15.7	9.7
Ni	1.6	1.6	N.R.	0.1	0.9	1.0	3.6
Cu	2.2	2.4	N.R.	0.1	2.1	0.6	2.8
B	N.R.	0.5	0.7	0.5	N.R.	0.1	18.5
Si	N.R.	0.2	7.2	5.0	N.R.	0.6	7.9
Al	N.R.	0.3	4.2	0.1	N.R.	0.2	3.1
Ag	N.R.	14.8	N.R.	0.1	N.R.	2.0	N.D.

N.R. = Not Reported, N.D. = Not Detected

clides (Table I); also there was no intergranular attack of the base
metal, i.e. no penetration of cesium activity into the base metal
(5).

Subsequent analyses on other sections of the H8 and B8 leadscrews
generally confirmed the presence of two distinct layers covering the
base metal, i.e., an outer layer of brushoff debris (LAD) and an
adherent inner layer (AD). Also the average ^{137}Cs surface activities
were comparable to that observed on H8-6, i.e., H8-7/13 = 889
µCi/cm^2, H8-7/16 = 792 µCi/cm^2 and B8-1 = 553 µCi/cm^2 (7). Chemical
examinations indicated both the LAD and AD had various elemental
compositions (Table III). Furthermore, physical and metallographic
analyses (Rockwell hardness measurements and grain structure morphology) provided the basis for temperature estimates for upper and
lower sections of the H8 and B8 leadscrews (Figure 1; 7).

The analytical results from the initial detailed examinations on the
H8 leadscrew included SEM examination, microchemical and X-ray
diffraction analyses. The SEM examinations confirm the optical
microscopy results of the number of surface deposit layers and confirm the lack of intergranular base metal attack. The microchemical
analyses involved Auger microprobe and electron microprobe analyses.
The Auger results indicate all three regions consist primarily of Fe,
Cr, Ni, and O (Table IV). These analyses also disclose metallic
regions (primarily silver and indium) embedded in the LAD/AD layer
(Figure 5). The electron microprobe results also confirm no cesium
intergranular attack of the base metal. Figure 5 also provides
microprobe X-ray line scans for 12 elements on metallographic sample
3-3A (5).

Although the microprobe line scans results are not quantitative,
general compositional trends and elemental associations are observed.
It is these line scans that provided evidence for the hypothesis that
a tenacious film rather than uniform diffusion or intergranular
attack probably accounts for the observed adherent cesium activity.

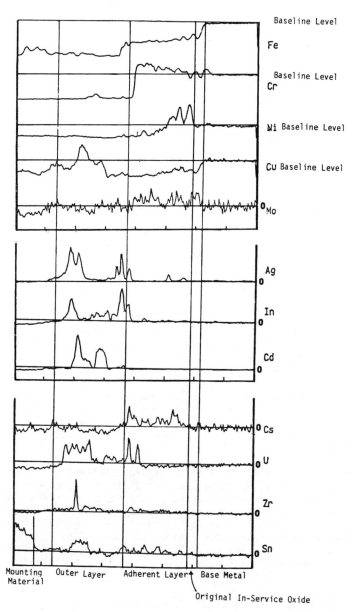

Figure 5   Microprobe Line Scan Results For Metallographic
Mounting 3-3A From The H8-5 Leadscrew Section. "Reproduced with
permission from Ref. 5. Copyright 1984, 'Electric Power
Research Institute'"

Table III.   Comparison of Analyses Reported for H8-7 and B8-1
for Brushoff and Inner Layers

	--------LAD[a]--------		--------AD[b]---------	
Element	H8-7 Wt.%	B8-7 Wt.%	H8-7/15 Wt.%	B8-1/7 Wt.%
Fe	37.0	30.0	41.6	16.4
Zr	0.4	2.0	0.2	1.8
U	N.D.	1.0	N.R.	0.2
Cr	22.0	11.0	9.7	5.0
Ni	0.1	1.0	3.6	2.4
Cu	0.1	0.1	2.8	1.5
Si	5.0	0.1	7.9	2.6[c]
Al	0.1	0.1	3.1	1.1
Ag	0.1	0.1	N.D.	14.7

N.D. = Not Detected, N. R. = Not Reported
a) Analysis of brushoff debris for entire section. Data tables
7 & 16 of (7)
b) Analysis of dissolution solutions for section specimens. Data
tables 8 & 17 of (7)
c) Serial solutions 7c + 7d were utilized in evaluating Wt.%.

Source: Adapted from Ref. 7.

Note for example, the relative decrease in Cr level and apparent rise
in Cs level at the LAD/AD interface. Consequently, additional high
resolution SEM characterizations studies were undertaken to examine
this interface (8).

The results show a Dense Region (DR) of 1 to 5 micrometers existed in
the outer region of the AD layer (Figure 6). This dense region has a
cermet type structure, a fact consistent with the "glassy" appearance
and the observed impervious/tenacious characteristics of the adherent
cesium activity reported from dissolution tests. The composition of
the dense region was found to be non-uniform with the principal
constituents detected in two adjacent areas being Cr, Fe, Ni, and Si,
Cu, Sn respectively. Surface analysis indicated a Si peak accom-
panied by a corresponding decrease in the Cr level at the outer edge
of the AD (Figure 6). Additional analysis indicate the presence of
Cs peaks at the same spatial positions as the Si peaks (Figure 7).
The observation of cesium and silicon intensities peaking at the same
spatial positions for stainless steel systems is consistent with
current laboratory experimental studies in the high temperature
fission product and transport program (11).

CRDM Leadscrew Support Tube and RTD Thermowell

CRDM H8 Leadscrew Support Tube. The examinations of a 3.5 inch
section removed from the bottom of the Leadscrew Support Tube (LST)
involved separation and dissolution of surface deposits and subsequent analyses that included SIMS, ESCA, ICAP and XRD techniques (9).
Given the close proximity of the section samples from the H8 LST and
the H8 leadscrew, it was anticipated that the LST analysis results
would be similar to those reported for the H8 leadscrew sections and
indeed the results are generally consistent.

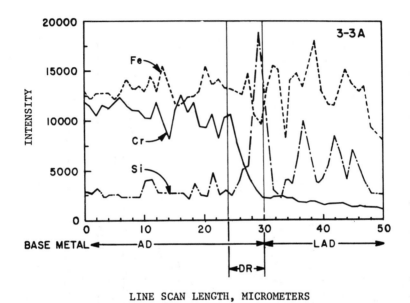

LINE SCAN LENGTH, MICROMETERS

Figure 6   WDX Line Scan Results For Metallographic Mounting 3-3A
From The H8-5 Leadscrew.

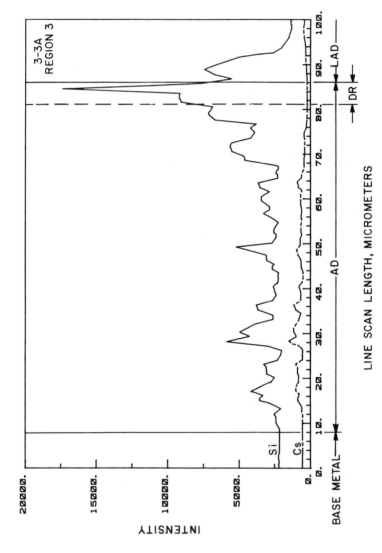

Figure 7 WDX Line Scan Results For Metallographic Mounting 3-3A
From The H8-5 Leadscrew.

Table IV. Auger Analysis Results for Base Metal
and Adherent Layers (5)

Sample	Location	Area	O Wt.%	Cr Wt.%	Fe Wt.%	Ni Wt.%	Cu Wt.%
1-3A	Thread Top	LAD	15	20	58	7	0
		LAD	12	20	66	2	0
		LAD	14	23	63	0	0
		AD	11	18	65	6	0
		AD	9	17	52	10	12
		BM	1	12	84	4	0
		M/O	6	19	70	5	0
5-3A	Top Root	LAD	8	15	42	15	20
		AD	10	20	55	12	3
		M/O	1	11	84	3	2
		BM	1	12	83	4	1
		AD	14	23	54	8	0
		LAD	12	19	56	12	1
		LAD	10	16	60	11	2
		M/O	1	11	84	3	1
	Thread Top	M/O	9	16	65	8	2
		M/O	9	17	72	2	0
		M/O	8	11	69	12	0
		BM	1	10	82	4	2
		BM	1	10	83	4	3
		BM	3	11	80	5	1
		BM	1	11	84	4	1
		M/O	13	18	63	6	0

The area designations BM (Base Metal), M/O (Metal Oxide adjacent to
BM), AD (Adherent Surface Deposits adjacent to M/O), and LAD (Loosely
Adherent surface deposits adjacent to AD) are general designations
that have specific reference locations which are correlated with the
photographs in (5).

Surfaces on both the Outside (OD) and Inside (ID) of the LST were
examined. The samples had approximately equal OD and ID surface
areas, with a total contaminated surface area per sample in the range
of 0.1 to 0.2 cm^2. The principal findings are as follows:
(1) Metallographic examination indicated the presence of two layers,
a LAD and a AD, with the ID-LAD layer appearing to differ from the
OD-LAD layer in thickness and porosity. The entire outside surface
appeared black with some yellow-orange deposits and the flat bottom
of the LST contained some spherical metallic particles up to 1
millimeter in size.

(2) ^{137}Cs and ^{134}Cs were the principal sources of gamma activity for
the LST with over 80% of this activity associated with the AD for the
OD surface. The reported ^{137}Cs surface activity was 1250 μCi/cm^2 for
the OD and 633 μCi/cm^2 for the ID. The ^{90}Sr activity was principally
associated with the LAD.

(3) Microstructural examination of the base metal (304 stainless
steel) indicated the LST did not experience temperatures, during the
accident, greater than those involving manufacture, i.e., 950 $^{\circ}$F -
1350 $^{\circ}$F.

(4) Several section samples from the LST were subjected to sequential
dissolution steps initially performed on the first H8 leadscrew
section H8-6 (4) with essentially identical results, i.e., the most
aggressive dissolution step (HNO_3-HF) rapidly removing essentially
all of the adherent cesium activity. It was observed from these
various solution chemistry dissolution steps that the solutions were
functioning by undercutting the contaminated layer and leaching the
^{137}Cs activity. This observation was also made in conjunction with
the dissolution efforts on the H8-5 leadscrew section samples (5).

(5) Elemental analysis results for both the OD and ID LST LAD and AD
are generally consistent with the results from the leadscrews.

(6) The SIMS analysis results for Cr, Ni, and Si are consistent with
the line scans reported from the leadscrew analyses (5,8).

RTD Thermowell. A dual element resistance thermal detector (RTD) was
removed from the hot leg of the Once Through Steam Generator (OTSG) -
A (10). The RTD had been located in the OTSG-A "candy cane" (Figure
8) where it measured the temperature of the reactor coolant entering
the OTSG-A hot leg (normal temperature range of 520 to 620 $^{\circ}$F).
Consequently, the RTD sensed temperature changes during the accident.
The temperatures began increasing approximately 2 hours after reactor
scram, reached a maximum of 800 $^{\circ}$F at approximately 3 hours and then
oscillated around 700 $^{\circ}$F for about 10 hours before decreasing. The
thermowell was removed in April 1984 for laboratory examination (10).
Visible inspection revealed no loose debris. The tip was orange and
the remaining surfaces brown or brown/black; the average ^{137}Cs
surface activity was 20 μCi/cm^2. The initial dissolution steps
carried out on the first H8 leadscrew section were also initially
performed on the thermowell and resulted in only partial removal of
the adherent cesium activity (approximately 1/3). Dissolution of the
adherent cesium activity was effected with additional aggressive
solutions at elevated temperatures, i.e., 1N HCl used alone and
serially with 10 Wt.% HNO_3 - 0.1M HF at 200 $^{\circ}$F. These solutions were
analyzed and the principal elements detected were B (33.8 Wt.%) and
Fe (62.4 Wt.%). Other elements measured were U (2.5 Wt.%), Ag (0.6
Wt.%), Cu (0.4 Wt.%), Zr (0.3 Wt.%), with Cd and Te below detection
limits (10).

No detailed surface characterization studies were conducted; hence,
no elemental or radionuclide profile data are available for comparison with the leadscrew data. However, the reduced level of adherent
cesium retained by this RCS component sample relative to the leadscrew and leadscrew support tube samples (approximately a factor of
50) is a significant observation. It suggests that primary system
surfaces outside the reactor vessel retained only limited quantities
of cesium, a pertinent point from both a decontamination and source
term viewpoint.

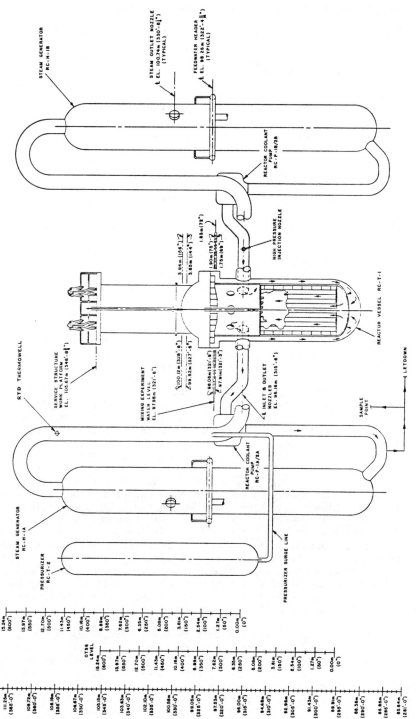

Figure 8 Schematic Of RCS Level Indication Versus RCS System Elevations Showing RTD Thermowell Location.

Cesium/Stainless Steel Interaction Laboratory Studies

Safety reviews for Light Water Reactors (LWR) have examined TMI-2
data and have recommended modified analyses of severe accidents
regarding fission product retention in the reactor coolant system
(12). Particularly relevant experimental studies regarding cesium
retention (adherence) on reactor material surfaces are underway at
Sandia National Laboratories (SNL) in their High Temperature Fission
Product Chemistry and Transport Program (11). Similar work is also
being carried out at AEE Winfrith Great Britain (13). The SNL
program was developed to investigate the chemistry and interactions
that might affect the transport of fission products from the fuel
into the reactor containment. SNL experiments have been performed in
a steam facility in which coupons of 304 stainless steel or Inconel
600 alloys were exposed to CsOH or CsI vapors. The SNL experiments
generally involved introducing a gas phase cesium compound (i.e.
hydroxide or iodide) into a test chamber containing stainless steel
coupons whose atmosphere and temperature were varied for different
experiments. Two basic forms of cesium were noted with respect to
dissolution by water from the coupons: 1) removable and 2) adherent.
The coupons exposed to the CsOH and CsI were cross-sectioned and examined with microprobe methods and the oxide scale on the 304 stainless steel alloy was found to consist of a duplex layer. The outer
layer is primarily magnetite, Fe_3O_4 ($FeO \cdot Fe_2O_3$). The inner oxide
layer was composed of iron-chromium oxides, having a spinel-type
structure i.e., $FeCr_2O_4$, and a dispersed metallic phase whose major
components are nickel and iron. The relative extent of this duplex
oxide, i.e., its relative composition, is a function of the H_2/H_2O
ratio (oxygen chemical potential), temperature/time, etc.. It was
noted that silicon was found within the inner layer but not the outer
layer. Its distribution within the inner layer was not uniform and
hence may have accumulated at the oxide grain boundaries as silica.
Furthermore, the microprobe results also indicated that cesium was
non-uniformly distributed within the inner layer at the same spatial
locations as the silicon. A comparison of the cesium and silicon
distributions indicated a strong correlation. In addition, the
distribution of cesium did not correlate with the distribution of any
other element, thus suggesting a cesium + silicon compound, a cesium
silicate the most likely choice. SNL also determined the ratio of
silicon to cesium by plotting the silica content for various local
areas versus their cesium content, as shown in Figure 9. It
should also be noted from Figure 9 that the data cluster about a
line that corresponds to a cesium to silicon ratio of 1:2 which
corresponds to the cesium silicate compound, $Cs_2O \cdot 4SiO_4$ or $Cs_2Si_4O_9$.
These data suggest that a particular compound is formed and imply
that the total amount of cesium that may be retained by an oxidized
stainless steel surface is directly proportional to the amount of
silica present in the surface oxide.

The SNL investigators found retained cesium to range approximately 3
$\mu g/cm^2$ to 15 $\mu g/cm^2$ depending, of course, on test parameters. The
corresponding equivalent ^{137}Cs data from Section 2.0 show general
agreement, i.e. H8-6 = 9.7 μg $^{137}Cs/cm^2$, H8-7/13 = 10.3 μg $^{137}Cs/cm^2$,
H8-7/16 = 9.2 μg $^{137}Cs/cm^2$, B8-1 = 6.4 μg $^{137}Cs/cm^2$, and LST = 14.4
μg $^{137}Cs/cm^2$. The total surface area of the upper plenum for TMI-2

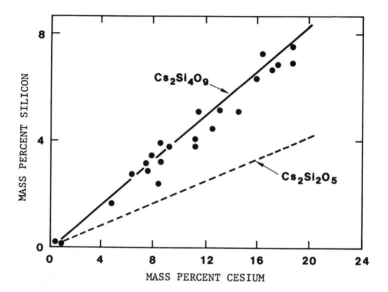

MASS PER CENT OF CESIUM AND OF SILICON FROM
NUREG/CR-3197 **PLOTTED TO SHOW THAT THE
PROBABLE REACTION SPECIES IS Cs$_2$Si$_4$O$_9$**

Figure 9   Correlation Of Silicon And Cesium Contents Of Local
Areas Within The Inner Oxide Layer Of A Cross-Sectioned 304
Stainless Steel Coupon. "Reproduced with permission from
Ref.11. Copyright 1984, 'Sandia National Laboratory'"

is approximately 10,309 ft^2 and if this entire area retains similar
cesium surface activity, then the total activity retained would
correspond to approximately 1% of the entire TMI-2 ^{137}Cs inventory
(8.18E5 Ci x 0.01 = 82,000 Ci). This calculation illustrates the
potential significance of surface retention from both a source term
and decontamination viewpoint. It should also be noted that the
experimental observations of cesium/silicon correlations (SNL laboratory and TMI-2) are consistent with materials utilized in engineering
fuel design to "retain" fission product cesium (14) and known mineral
forms, e.g., pollucite (15).

The weighting of pertinent parameters that effect the mechanism(s)
for cesium adsorption and retention on the TMI-2 internal surfaces
are illustrated by the Cs-137 surface activity as a function of
leadscrew length for H8, E9, B8, and the vessel plenum (Figure 10).
Given that three of the pertinent parameters effecting cesium adsorption and retention are temperature, oxygen chemical potential, and
oxide composition/structure then, the data presented in Figure 10
illustrating the similar response for the H8 leadscrew (17-4PH) and
the vessel plenum (304) as well as that for the E9 and B8 leadscrews
suggests that the oxygen chemical potential would appear to be the
most significant of these three parameters.

The laboratory experiments at UKAEE Winfrith (AEEW) produced qualitatively similar results. The exact mechanism and kinetics of formation of the retained cesium form have not been determined unequivo-
cally. The two groups studying these systems agree on the importance
of the inner oxide layer on the steel, but hypothesize different
mechanisms to explain the results. Differences may be due in part to
the different carrier gas and oxidation conditions (i.e. the carrier
gas was steam/H$_2$ in the relevant SNL experiments whereas the carrier
gas in the UKAEEW experiments was Ar, or Ar-4%H$_2$, or Ar-4%H$_2$-4%H$_2$O).
However, the point pertinent to this paper is that the adherent
cesium activity initially reported from simple solution chemistry
tests and subsequent surface characterization studies corresponds to
phenomena noted from published laboratory experiments that can be
partially quantified. Consequently, the adherent cesium question
initially raised would seem to be basically resolved even though full
understanding of the mechanism is not yet determined.

Conclusions

Several components have been removed from the TMI-2 reactor system.
Laboratory examinations were undertaken on samples from these components to characterize the decontamination requirements involved with
cleanup and to provide a data base to evaluate some of the physical
and chemical processes that occurred and conditions that existed
during the TMI-2 accident. The radionuclide ^{137}Cs was selected in
this review as its observed behavior bears directly on both the
decontamination questions and evaluation of chemical interaction
under accident conditions. Laboratory results on samples from the
components within the reactor vessel, i.e., leadscrews and leadscrew
support tube, show greater retained cesium activity than samples from
the components outside the reactor vessel, i.e., RTD thermowell,

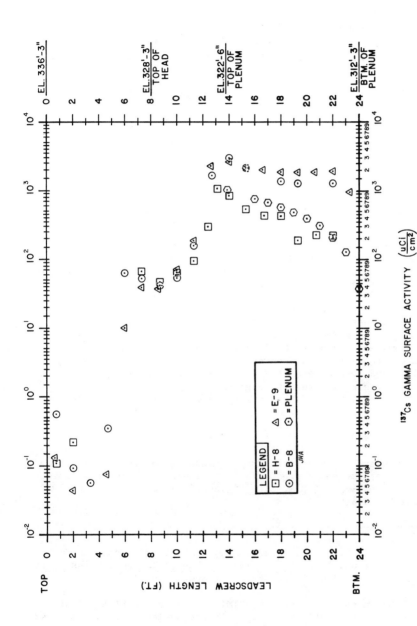

Figure 10 Cs-137 Gamma Surface Activity For H8, B8, E9 Leadscrews.

though the corresponding temperatures were approximately the same as
shown in Table V.

Table V.   Comparison of RCS Components Surface Activity and
           Determined/Measured Temperature

RCS Component	^{137}Cs Surface Activity (μCi/cm^2)	Determined/Measured Temperature ($^{\circ}$F)
H8-5	1100	N.R.
H8-6	840	N.R.
H8-7/13	889	800
H8-7/16	792	800
B8-1	553	900
LST	1250	<930
RTD	20	800
N.R. = Not Reported		

Since the temperatures of the upper plenum surfaces and the RTD
thermowell ranged around 800-900 $^{\circ}$F and yet the thermowell retained
less cesium, local atmosphere composition and resulting surface oxide
form likely played an important role.

Retention of cesium activity by the reactor vessel internal surfaces
is clearly evident from the surface activities observed for the
leadscrew and support tube components removed from TMI-2. These
observations are supported by laboratory experiments at SNL and AEEW.
These data are relevant to source term releases under accident
conditions and also suggest that decontamination of surfaces within
the reactor vessel may require aggressive techniques. However,
surfaces of components of the primary system which are outside the
reactor vessel may not have a significant adherent cesium activity.

Acknowledgments

The authors wish to express their gratitude to the referenced organizations who provided reproductions of figures from their documents
and to J. H. Aldinger (TMI-2 Site Engineering) for assistance in
preparing the graphics.

Literature Cited

1.   M. Rogovin, Director, Nuclear Regulatory Commission Special
     Inquiry Group, Three Mile Island, A Report To The Commissioners
     And To The Public, NUREG/CR-1250 (January 1980).

2.   D. E. Owen and M. R. Martin, "TMI-2 Core Examination: First
     Results", Transactions of the American Nuclear Society, Vol. 43,
     p.5 (November 14-18, 1982).

3.   J. A. Daniel, E. A. Schlomer, T. L. McVey, Analyses Of The H-8,
     B-8, & E-9 Leadscrews From The TMI-2 Reactor Vessel, SAI-83/1083

(31 August 1983), GPU Nuclear Corporation Data Report
TPO/TMI-097, Rev. 0, (November 1983).

4. K. J. Hofstetter, H. Loewenschuss, V. F. Baston, Data Report:
    Chemical Analyses and Test Results For Sections of The TMI-2 H8
    Leadscrew, GPU Nuclear Corporation TPO/TMI-103 (February 1984).

5. G. M. Bain and G. O. Hayner, Initial Examination of The Surface
    Layer of a 9-Inch Leadscrew Section Removed From Three Mile
    Island-2, EPRI Research Project 2056-2, NP-3407 (January 1984).

6. R. L. Clark, R. P. Allen, M. W. McCoy, TMI-2 Leadscrew Debris
    Pyrophoricity Study, GEND-INF-044 (August 1983).

7. K. Vinjamuri, D. W. Akers, and R. R. Robins, Draft Report:
    Examination of H8 and B8 Leadscrews From Three Mile Island Unit
    2 (TMI-2), EGG-TMI-6685 (October 1984).

8. V. F. Baston, K. J. Hofstetter, G. M. Bain, G. O. Hayner,
    "Initial Examination of Decontamination Barrier on TMI-2 Lead-
    screw", National Association of Corrosion Engineers Symposium,
    Corrosion 85 (March 25-29, 1985).

9. M. P. Failey, V. Pasupathi, M. P. Landow, M. J. Stenhouse, J.
    Ogden, and R. S. Denning, The Examination Of The Leadscrew
    Support Tube From Three Mile Island Reactor Unit-2, Final Report
    to Three Mile Island, Battelle Columbus Laboratories (to be
    published).

10. D. W. Akers, et.al., Analysis Of TMI-2 'A' Steam Generator Hot
    Leg Resistance Thermal Detector, GEND-INF-014 (to be published).

11. R. M. Elrick and R. A. Sallach, Reaction Between Some Cesium-Io-
    dine Compounds and The Reactor Material 304 Stainless Steel,
    Inconel 600, and Silver. Vol. I: Cesium Hydroxide Reactions,
    NUREG/CR-3197 (July 1984).

12. William Stratton, Chairman, "Report of The ANS Special Committee
    on Source Terms", Transactions of The American Nuclear Society,
    Vol. 47, p.18 (November 1984).

13. B. R. Bowsher, S. Dickinson, A. L. Nichols, J. S. Ogden, P. E.
    Potter, "Chemical Aspects of Fission Product Transport in The
    Primary Circuit of a Light Water Reactor", ANS Topical Meeting
    on Fission Product Behavior and Source Term Research at Snow-
    bird, Utah (July 1984).

14. R. Forthmann, H. Grubmeier, D. Stover, "Metallic Fission Product
    Retention of Coated Particles With Ceramic Kernel Additives",
    Nuclear Technology, Vol. 35, p.548, (September 1977).

15. Pollucite, $Cs_4Al_4Si_9O_{26}$ H_2O, a natural cesium aluminum silicate
    (see G. G. Hawley, Co-editor, The Condensed Chemical Dictionary,
    van Nostrand Reinhold Company (1977)).

RECEIVED July 16, 1985

8

Core Debris Chemistry and Fission Product Behavior

D. W. Akers

Idaho National Engineering Laboratory, EG&G Idaho, Inc., Idaho Falls, ID 83415

Examinations are being performed to acquire data on the
extent and nature of damage to the Three Mile Island
Unit 2 (TMI-2) core. Six samples were obtained from
the TMI core rubble bed in September 1983. Five of the
six samples are being examined by EG&G Idaho, Inc., to
acquire data on the postaccident condition of the core
and the behavior of radionuclides in the core region.
A description of the sampling and analysis methods is
presented, along with the results of chemical and radiochemical examinations. Correlations are performed with
the core structural composition and predicted fission
product concentrations and inventories. The core debris
is a mixture containing UO_2 fuel, zircaloy cladding,
and control materials (silver, indium, and cadmium);
poison rod materials; and structural materials. Radionuclide concentrations are similar at all locations.
The principal radionuclides measured are strontium-90,
iodine-129, cesium-137, and cerium-144.

On March 28, 1979, an accident occurred at Three Mile Island that
resulted in severe damage to the core of the Unit 2 pressurized water
reactor. Examination of accessible reactor systems was begun shortly
after the accident to provide data on the causes and effects of
severe core damage accidents and to assist General Public Utilities
Nuclear Corp. in defueling the reactor.

Figure 1 shows the current condition of the TMI-2 core, as determined from closed-circuit television, core topography scanning,
and rubble bed probing examinations. It is estimated that approximately 20% of the total core mass is an upper layer of rubblized
debris supported by a hard crust. Approximately 65% is located between the debris bed and the eliptical flow distributor, and approximately 10-20% has relocated to the reactor vessel lower plenum. The
reactor core was sampled in September 1983; the initial six samples
were particulate debris removed from the rubble bed.

Examination of the core debris samples was performed to acquire
data on the extent and nature of core damage and the postaccident

0097-6156/86/0293-0146$06.50/0
© 1986 American Chemical Society

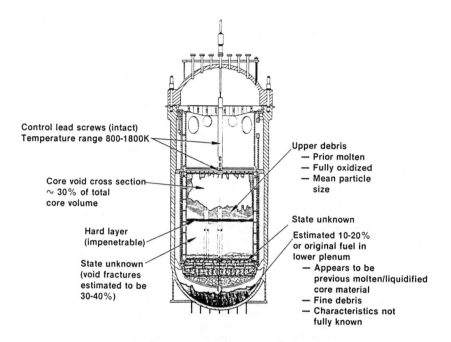

Control lead screws (intact)
Temperature range 800-1800K

Core void cross section
~ 30% of total
core volume

Hard layer
(impenetrable)

State unknown
(void fractures
estimated to be
30-40%)

Upper debris
— Prior molten
— Fully oxidized
— Mean particle
  size

State unknown

Estimated 10-20%
or original fuel in
lower plenum
— Appears to be
  previous molten/liquidified
  core material
— Fine debris
— Characteristics not
  fully known

Figure 1. Best estimate of end-state TMI-2 core conditions
following the incident.

condition of the core. The principal objectives of the core debris examinations are to determine:

- The physical form of the particulate core debris (particle size, shape, morphology, origin, etc.)

- The origin, chemical forms, and composition of the debris (fuel, cladding, control material, structural material, reaction product, etc.)

- The identity and quantity of fission products retained in the debris.

Examinations performed, but not discussed in this paper, include the pyrophoricity of the debris, ferromagnetic material content, the physical behavior of the debris as it relates to defueling the reactor (1,2), and the fuel behavior analyses, which have not been completed.

The six samples of particulate debris from within the TMI-2 rubble bed were obtained by lowering sampling devices through control rod leadscrew openings at two locations in the TMI-2 core, H8 (mid-core) and E9 (mid-radius). Two different sampling devices were used to extract samples from the rubble bed. One was a clamshell-type tool used to take surface samples; the other was a rotating-tube device with doors on each side of the tube. Figure 2 shows the TMI-2 core debris sampling schematic and the two sampling tools.

The core debris samples were obtained from three depths in the rubble bed: surface, 3 in., and 22 in. After their removal, the six samples were shipped to EG&G Idaho at the Idaho National Engineering Laboratory (INEL). The sample obtained at the H8 location, 3 in. into the debris bed, was subsequently shipped to the Babcock & Wilcox (B&W) Lynchburg Research Center for examination (3). The five remaining samples are being examined at the INEL. It should be noted that the samples have been immersed in reactor coolant for more than 5 years and that this may affect the results of the analysis.

This paper describes (a) the sampling methods used to obtain portions of the bulk samples for analysis; (b) the examination techniques used for analysis; (c) the bulk samples, including particle size analysis; and (d) the chemical and radiochemical analysis results obtained to date. The energy dispersive X-ray (EDX) analysis data and scanning auger spectroscopy (SAS) data are referred to only briefly, because these examinations have not been completed.

Sampling and Measurement Methods

Analysis of the core debris samples posed two related problems caused by the high radiation fields (2-36 Rad/h) associated with the bulk samples: (a) most chemical and radiochemical analysis techniques used outside a hot cell are limited to relatively low (<400 mRad/h) radiation fields, and (b) by reducing the quantity of material analyzed to thus reduce the radiation field, the small sample analyzed may not be representative of the bulk sample. Therefore, a sampling and analysis approach was developed which allows for the analysis of small samples for characterization of individual particle types and for representative analysis of the

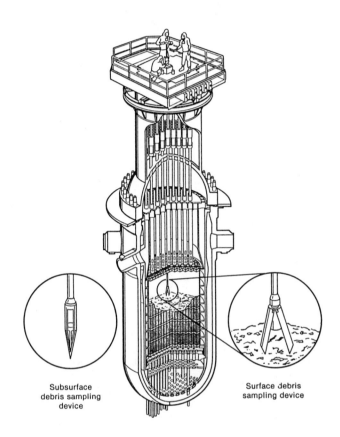

Subsurface
debris sampling
device

Surface debris
sampling device

Figure 2. TMI-2 core debris sampling schematic.

bulk sample. Each bulk core debris sample was photographed, weighed, and then sieved into nine particle size groups, ranging from 4000 μm to <30μm. Individual particles were selected from particle size groups >1000 μm, and aliquots were removed from size groups <1000 μm. The number of particles and aliquots removed from each bulk sample ranged from 5 to 16. Two of the core debris samples (4 and 5) were not subjected to a full particle size analysis because they were composed principally of large (>1000 μm) particles.

Portions (<400 mRad/h) from the individual particles and sample aliquots were then removed and prepared for chemical and radiochemical analysis. The remainder of each core debris sample was retained for metallurgical, EDX, and SAS analysis.

To characterize a core debris bulk sample, ∿33% of each bulk particle size group was weighed, recombined, dissolved, and analyzed. Figure 3 shows the general analysis scheme for all samples. The particle size analysis was performed to determine particle size distribution and whether there is a correlation between particle size and chemical or radiochemical behavior.

As indicated in Figure 3, gamma spectroscopy and fissile/fertile material analyses were performed on the particle and aliquot sample fractions, which were then dissolved for iodine-129, strontium-90, and elemental analysis. The method used to dissolve the particle and aliquot sample fractions was a potassium bisulfate fusion in a closed system. This method was used so that quantitative analysis for iodine-129 could be performed. Both carrier iodine and tracer iodine-131 for chemical yield determination were included in the dissolution.

The bulk sample analysis was performed by recombining 1/3 of each of the particle size groups and subsequently dissolving the sample using a sequential dissolution in the following order: (1) 6 $\underline{M}$ HNO$_3$, (2) 3 $\underline{M}$ HNO$_3$ plus 1 $\underline{M}$ HF, and (3) aqua regia. Analyses were then performed on a sample of the combined solutions. It was not possible to analyze these samples for volatiles (e.g., iodine-129), because the chemical yield was not measurable due to the sequential dissolution. The analytical methods used for the core debris analyses are listed in Table I, with a description of the sample type analyzed and the measured parameter. Exact details of the analytical methods and associated uncertainties are yet to be published.(4)

Bulk Sample Descriptions

At EG&G Idaho, the five bulk samples (1, 3, 4, 5, and 6) were unpackaged, visually inspected, and photographed. A summary of the bulk examination, including gross radiation fields for each sample, is presented in Table II.

Each bulk sample was weighed. Then, a tap bulk density measurement (i.e., approximate volume and weight) was performed on Samples 1, 3, and 6, which are most representative of the debris bed. The tap density ranged from 3.5 to 3.8 g/cm^3 for the three samples. The tap density for Samples 4 and 5 was not measured, because these samples were composed primarily of large particles.

Sample 1, the surface sample from the H8 location, was obtained using a clamshell sampler (see Figure 2). The sample contained approximately 71 g of very black debris, with particle sizes ranging

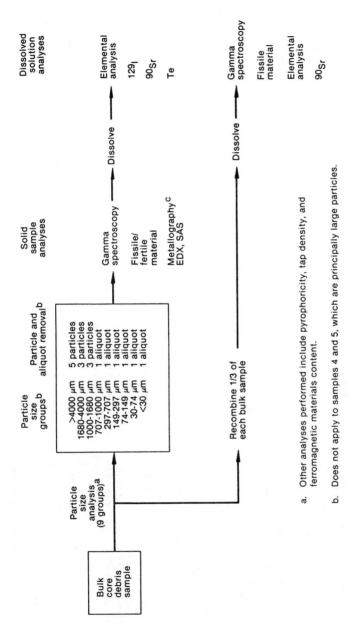

Figure 3. TMI-2 core debris analysis scheme.

a. Other analyses performed include pyrophoricity, tap density, and ferromagnetic materials content.

b. Does not apply to samples 4 and 5, which are principally large particles.

c. Metallography, SEM/EDX, and SAS were performed on the portions of large particles not used for the chemical and radiochemical examination.

TABLE I. ANALYTICAL METHODS AND MEASURED PARAMETERS

	Analytical Method	Sample Analyzed	Measured Parameter
1.	Gamma spectroscopy	Solid or liquid sample fractions (mass attenuation correction performed)	Concentrations of Co-60, Ru-106, Ag-110m, Sb-125, Cs-134, Cs-137, Ce-144, Eu-154, and Eu-155
2.	Neutron activation with subsequent delayed neutron analysis	Solid sample analyzed for fissile/fertile material	Principal components analyzed are U-235 (fissile) and U-238 (fertile)
3.	Inductively coupled plasma spectroscopy (ICP)	Dissolved samples by ICP spectroscopy	Elements analyzed for: Ag, Al, B, Cd, Cr, Cu, Fe, Gd, In, Mn, Mo, Ni, Nb, Si, Sn, Te, U, and Zr
4.	Radiochemical (Sr-90)	Dissolved samples by chemical separation with subsequent beta counting for Y-90.	Concentration of Sr-90
5.	Radiochemical (I-129)	Volatile fraction of dissolved sample with carrier and I-131 for chemical yield determination	Concentration of I-129
6.	Metallography scanning electron microscopy (SEM) SAS, EDX	Solid sample	Determine basic structures present, approximate composition, interaction

TABLE II.   SUMMARY OF VISUAL PHOTOGRAPHIC EXAMINATION AND GROSS RADIATION LEVELS

Sample Number	Sampler Type	TMI-2 Core Location	Location of Sample in Rubble Bed	Gamma Radiation Level at 1-in.[a] (Rad/h)	Visual Characteristics[b]
1	Clamshell	H8	Surface	16	A pile of very black, damp debris with a fairly wide range of particle sizes (dimensions ranging from 0.2 to 1.3 cm); several rounded surfaces; sporadic rust color throughout.
3	Rotating Tube	H8	22 in. into debris bed	36	Very black debris, slightly damp, wide range of particle sizes (dimensions 0.2 to 0.6 cm), small chunks to fine debris; similar to Sample 1.
4	Clamshell	E9	Surface (surface)	3	Thirteen major chunks, dry, black with rust colored sides, basically sharp edges with one or two chunks having rounded edges; dimensions ranging from 0.2 to 1.0 cm
5	Rotating Tube	E9	3 in. into debris bed	18	Similar to Sample 4 with the following distinctions: many more pieces; greater size range (1/16 to 3/8 in.); some surfaces more reflective. Again, very dry.
6	Rotating Tube	E9	22 in. into debris bed	36	Very black debris, small chunks to fine debris, slightly damp, some pieces blackish gray. A couple of pieces resembled metal shards similar to Sample 3.

a. Radiation levels measured using a teletector probe on the exposed sample.

b. All samples appeared to be of a loose granular nature with some sharp or rounded edges.

from 30 μm to greater than 4000 μm; the majority of the material
was greater than 1000 μm. A unique particle in Sample 1 is a large
fuel rod remnant, approximately 19,000 μm long, consisting of zirconium cladding with pieces of fuel attached (Figure 4). The outer
surface of the cladding and all fracture surfaces appear smooth. In
general, the particles within the debris have smooth surfaces and
rounded corners.

Sample 3, the 22-in.-deep sample from the H8 position, was obtained using a rotating tube sampler. The material was stratified
within the sampler, with the larger particles toward the top and the
finer particles nearer the bottom. This stratification is likely to
have occurred during shipment. The particle sizes in this sample
ranged from <30 μm to greater than 4000 μm, as shown in
Table III, with a majority of the material (90 wt%) greater than
1000 μm in size. This sample contained several particles that
appeared to be fractured fuel pellets.

Sample 4, the E9 surface sample, consisted of 13 large-sized
particles (ranging from 1000 μm to approximately 4000 μm). All
pieces have the appearance of fractured fuel pellets.

Sample 5, the E9 3-in.-deep sample, contained almost all large
(>1000 μm) particles and appears to be similar to Sample 4.

Sample 6, the E9 22-in.-deep sample, is very similar in quantity and appearance to Sample 3, the H8 22-in.-deep sample. In
general, 80% or more of the debris examined (by weight) was larger
than 1000 μm.

The particle size distribution analysis was done by sieving the
bulk samples into a number of progressively smaller particle-sized
groups (nine for most samples). The results of these analyses are
shown in Table III. Both wet (freon wash) and dry sieving techniques were employed. For the larger particle size fraction (i.e.,
>1000 μm), dry hand-agitated sieving was used. For the size
groups <1000 μm, wet sieving (i.e., freon wash) was used to
reduce suspension of the smaller size particles. The freon wash was
used because it did not react chemically with the core debris
materials.

Results of Chemical Analysis

Analytical results principally addressed in this paper are those
obtained by Inductively Coupled Plasma (ICP) spectroscopy, because
the examinations by other techniques are still being performed.
Complete results will be included in the final report on the core
debris analysis (5). ICP analysis was performed on 65 particle and
aliquot sample fractions obtained from the five grab samples.
Table IV lists the initial composition and total core weight of the
principal core components (6) and identifies the elements analyzed.
The elements were chosen to characterize the three principal groups
of materials present in the core: (a) the uranium fuel and zircaloy
cladding material, (b) the silver-indium-cadmium (Ag-In-Cd) control
rod and burnable poison rod materials, and (c) the structural
materials (stainless steel and Inconel).

Table V lists the average elemental concentrations measured in
the five core debris samples. The principal components measured in
all samples analyzed are uranium and zirconium. Silver, aluminum,
chromium, iron, manganese, and nickel components of control rods and

Figure 4. Fuel rod segment from surface of debris bed at H8 location (Sample 1A).

TABLE III. RESULTS OF PARTICLE SIZE ANALYSIS

Particle Size Range (μm)	Sample No. 1[a] (g)	(%)	Sample No. 3 (g)	(%)	Sample No. 4[b] (g)	Sample No. 5[c] (g)	(%)	Sample No. 6 (g)	(%)	Sample No. 6[d] (g)
>4000	12.62	18.4	63.75	42.9		69.57	77.1	57.99	42.0	0
1680 to 4000	27.82	40.6	51.45	34.7		13.96	15.5	49.39	35.8	0.39
1000 to 1680	15.64	22.8	19.19	12.9		6.25	6.9	13.88	10.1	0.30
>1000		81.8		90.5			99.5		87.9	
<1000						0.44	0.49			
707 to 1000	7.80	11.4	5.49	3.7				8.93	6.5	0.25
297 to 707	3.20	4.7	6.34	4.3				5.99	4.3	0.19
149 to 297	0.87	1.3	1.27	0.86				0.97	0.70	0.025
74 to 149	0.44	0.64	0.77	0.52				0.67	0.48	0.024
30 to 74	0.17	0.25	0.18	0.12				0.22	0.16	NA
<30	NA[e]	—	0.013	0.01				NA	-0-	NA
Summed wt	68.56		148.45			90.22		138.04		1.178
Initial wt	70.88		152.71		16.59	90.96		140.73		
Loss	2.32	3.3	4.26	2.8		0.74	0.8	2.69	1.9	

a. Sample numbers shown correspond to sample numbers listed in Table I.

b. Sieving was not done. Sample consisted only of large pieces.

c. Sieving was limited to four sizes. Sample consisted mostly of large pieces.

d. Ferromagnetic sample weights (g). These samples are a subset of their respective weight fractions for Sample No. 6.

e. None detected (not measurable).

f. The loss fraction defines an uncertainty in the quantity of material present: however, the loss fraction particle size distribution is not known.

TABLE IV. TMI-2 REACTOR CORE COMPOSITION

Material	Element	Weight %	Material	Element	Weight %
UO_2 (93 050 kg)	U-235[a]	2.265	Inconel-718 (1 211 kg)	Ni[a]	51.900
	U-238[a]	85.882		Cr[a]	19.000
	O	11.853		Fe[a]	18.000
				Nb	5.553
				Mo	3.000
Zircaloy-4 (23 029 kg)	Zr[a]	97.907		Ti	0.800
	Sn[a]	1.60		Al[a]	0.600
	Fe[a]	0.225		Co	0.470
	Cr[a]	0.125		Si[a]	0.200
	O	0.095		Mn[a]	0.200
	C	0.0120		N	0.130
	N	0.0080		Cu[a]	0.100
	Hf	0.0078		C	0.040
	S	0.0035		S	0.007
	Al[a]	0.0024			
	Ti	0.0020			
	V	0.0020			
	Mn[a]	0.0020			
	Ni[a]	0.0020	ZrO_2	Zr[a]	74.0
	Cu[a]	0.0020	(331 kg)	O	26.0
	W	0.0020			
	H	0.0013			
	Co	0.0010	Ag-In-Cd	Ag[a]	80.0
	B[a]	0.000033	(2 749 kg)	In[a]	15.0
	Cd[a]	0.000025		Cd[a]	5.0
	U[a]	0.000020	$B_4C-Al_2O_3$	Al[a]	34.33
Type 304 SS (676 kg)	Fe[a]	68.635	(626 kg)	O	30.53
	Cr[a]	19.000		B	27.50
Unidentified SS (3 960 kg)	Ni[a]	9.000		C	7.64
	Mn[a]	2.000			
	Si[a]	1.000			
	N	0.130	$Gd_2O_3-UO_2$	Gd[a]	10.27
	C	0.080	(131.5 kg)	U	77.72
	Co	0.080		O	12.01
	P	0.045			
	S	0.030			

a. Elements for which ICP analysis was performed.

TABLE V.   SUMMARY OF

Element[a]	H8 Surface (# 1) Total Analyses = 16		H8-22 In. (#3) Total Analyses = 15	
	Average Concentration	Number of Analyses	Average Concentration	Number of Analyses
Ag	2.1	14	2.0	12
Al	6.3	15	5.6	10
Cd	2.1	7	9.9 E-1	8
Cr	5.6	15	5.8	12
Cu	8.1 E-1	8	8.0 E-1	4
Fe	1.2 E+1	15	1.0 E+1	13
Gd	0	0	7.0 E-1	5
In	6.6	5	4.1	5
Mn	4.8 E-1	15	3.7 E-1	13
Mo	0	0	1.3	4
Ni	7.2	13	5.3	10
Nb	1.4	1	1.5	2
Si	6.5	14	7.6	15
Sn	2.6 E+1	8	8.5	6
Te	5.2	5	10	1
U	4.80 E+2	15	5.84 E+2	14
Zr	2.22 E+2	16	2.10 E+2	15

a.   Boron is not listed because it was measurable in few samples.

1ENTAL ANALYSIS RESULTS
;m)

E9 Surface (#4) Total Analyses = 5		E9-2 In. (#5) Total Analyses = 11		E9-22 In. (36) Total Analyses = 16	
verage entration	Number of Analyses	Average Concentration	Number of Analyses	Average Concentration	Number of Analyses
9.2 E-1	4	1.3	11	9.3	16
9.3	5	3.1	10	5.8	13
5.2 E-1	2	1.7 E-1	3	7.5 E-1	13
3.9	4	1.5	10	1.2 E+1	10
1.6	1	3.1 E-1	3	2.8 E-1	5
5.5	4	2.3	8	1.0 E+1	13
9.2 E-1	1	8.4 E-1	2	8.1 E-1	8
2.1 E+1	1	7.8	4	3.9	9
1.3 E-1	2	3.1	7	7.4 E-1	11
4.8	1	1.9	3	5.8 E-1	4
3.9	3	4.0	7	1.3 E+1	9
0	0	0	0	1.3	6
1.7 E+1	5	1.0 E+1	11	4.2	16
11	1	8.9	8	9.3	8
4.6	3	3.0	4	1.3	1
8.32 E+2	5	7.63 E+2	11	4.66 E+2	13
4.26 E+1	5	8.77 E+1	11	3.30 E+2	16

structural materials were measurable in most samples, indicating that the core debris samples analyzed were composites of the several core components.

Uranium and Zirconium. The core was originally composed of 82,000 kg uranium and 22,500 kg zirconium (Zr/U ratio ∿0.247). For comparison, the average Zr/U ratios for the samples analyzed were calculated based on particle and aliquot analysis to determine the fraction of zirconium and uranium present in each sample. These ratios are as follows: #1 (0.46), # 3 (0.36), #4 (0.05), #5 (0.11), and #6 (0.71). However, it was determined that the average Zr/U ratio obtained from the individual particle and aliquot samples provides a misleading indication about the average ratio of the bulk sample because the particles chosen for analysis were not representative of the debris bed. Some particles were chosen because they appeared to be intact fuel, zircaloy cladding, or structural material. The average Zr/U ratio for the core debris samples will be obtained from the analysis of the dissolved bulk samples.

Table VI shows the Zr/U ratios for all individual particles and sample aliquots analyzed. For the large size particle groups (>1000 μm), ratios range from 4.6 E-3, mostly uranium, to 13, mostly zirconium. For the smaller particle sizes, a more consistent pattern is indicated, with many particles having Zr/U ratios of 0.2 to 0.3, similar to the original core ratio. Some of the particles and aliquots measured have Zr/U ratios similar to the ratio of the total, intact amount of zirconium to the total amount of uranium in the core [i.e., the core ratio is similar to the ratio in a fuel rod (0.23)].(7) This indicates a homogenous mixing of the two elements, because the ratio was similar for large particles and sample aliquots.

There is a trend toward an increasing fraction of zirconium as particle sizes get smaller to approximately a 1:1 Zr/U weight ratio for the <30 μm size group, which is equivalent to an atom ratio of ∿2.6:1. Similarity in behavior at the H8 and E9 locations at 22 in. (i.e., concentrations and particle size distribution) indicates that both locations in the core were probably subjected to similar environments during and after the accident.

Control Rod Materials (Ag-In-Cd). The TMI-2 core had 61 full-length and 8 part-length control rod assemblies (16 rods per assembly), containing by weight 80% silver, 15% indium, and 5% cadmium (8), with a total weight of 2749 kg. H8 and E9 are both control rod assembly locations. From Table V, it is apparent that silver is measurable in most samples. The concentration range is from ∿1-2 mg silver per gram of debris, with the exception of Sample 6 (E9 at 22 in.). In the control rod assembly, silver accounts for ∿4.8% of the assembly by weight (8), a factor of 24 times greater than measured concentrations. Also, the silver in the control rods is concentrated at specific locations in the assembly. Therefore, much of the silver has been relocated. The melting point for silver is 1234 K, which is below the estimated temperature of the core during the accident (5). For sample 6 (E9 at 22 in.), a general increase in silver concentration is present for the smaller particle size fractions (<707 μm) to approximately 27 mg silver per gram of debris for the 30-74 μm size fraction. A corresponding increase

in the fraction of zirconium is also apparent from Table VI. The mechanisms that may have caused this behavior are to be evaluated further (5).
The data indicate that silver has been evenly distributed at the sampled locations (1-2 mg/g) with some localized accumulations. The higher concentrations at the smaller particle sizes suggest that silver may be present primarily as relatively small nodules or localized accumulations. Assuming a total core mass of ~1.25 E5 kg (6), an evenly distributed concentration of 2 mg/g of silver would be equivalent to 11% of the core inventory; however, an average concentration of 27 mg/g would be equivalent to 150% of the inventory. Other analyses, the H8 and E9 leadscrew analysis data (9), indicate that <2% of the core inventory of silver was found in the upper plenum. The absence of large fractions of the silver inventory in the upper plenum and core debris suggests that the silver has melted and relocated to regions lower in the reactor vessel.
Indium was measurable in fewer particle and aliquot sample fractions than was silver, as indicated in Table V, but the concentrations are generally higher, 3-8 mg of indium per gram of debris. The fewer number of samples where indium was measurable may be due to the ICP detection limit for indium, which is approximately a factor of four higher than the limit for silver. However, if the measured concentrations are averaged over all samples analyzed (64), the average indium concentration in the debris bed samples is ~2.2 mg/g, which is equivalent to 67% of the core inventory if extrapolated to the total mass of the core materials (1.25 E5 kg).
Cadmium was measurable in a larger fraction of the total number of samples than indium; but, whereas the samples with measurable amounts of indium were distributed through all particle size ranges, cadmium was measurable principally on particle size fractions <1000 μm. The smallest size fractions have the highest cadmium concentration (2.2-4.8 mg/g). If the cadmium sample concentrations are averaged over the 64 samples analyzed, the concentration of cadmium in the core is ~0.5 mg of cadmium per gram of debris, which is equivalent to ~49% of the core inventory of cadmium if extrapolated to the mass of the core.
The fractions of the core inventory of the control rod materials retained in the core, based on the analysis of individual particles, are estimates that have a high degree of associated uncertainty. Bulk samples (~50 g) have been dissolved and are being analyzed for elemental content to provide a more accurate indication of the quantity of control rod material retained in the bulk debris.

Poison Rod Materials. The poison rod materials measured were boron, aluminum, and gadolinium. Boron was measured in most samples and may have been deposited from the reactor coolant. The principal source of aluminum in the core is the $B_4C-Al_2O_3$ rods, which are ~34% aluminum. The data indicate that aluminum is widely dispersed in the core debris with little or no measurable boron present, indicating that these materials were separated during the accident. The melting point for $Al_2O_3 + B_4C$ is approximately 2400 K (8), indicating that this temperature had been exceeded. The average concentration of aluminum in the debris is ~4.8 mg/per gram of debris, as compared to the inventory of 1.7 mg/g in the

TABLE VI.   ZIRCONIUM-TO-URANIUM (ZR/U) RATIOS[a]

Particle Size	Sample Number				
	H8 Surface Number 1	H8-22 in. Number 3	E9 Surface Number 4	E9-22 in. Number 5	E9-22 in. Number 6
>4000 µm	2.7	6.0 E-2	1.4 E-1	1.4 E-1	4.2 E-1
	1.5 E-2	--b	3.1 E-2	4.8 E-2	--b
	1.5 E-1	--b	2.3 E-2	6.8 E-2	1.3 E+1
	6.3 E-2	1.9 E-2	8.3 E-3	4.6 E-3	1.0 E-2
	3.7	3.2 E-2	2.9 E-2	4.2 E-1	--b
1680-4000 µm	2.4 E-1	5.0 E-2	--c	5.1 E-3	--b
	2.4 E-1	1.4 E-1		--b	5.5 E-3
	7.8 E-1	2.2 E-1		3.7 E-2	5.8 E-3
1000-1680 µm particles	--b	1.5 E-1		4.9 E-1	1.9 E-1
	2.3 E-1	2.4 E-1		5.5 E-3	4.2
	2.3 E-1	1.9 E-1		7.9 E-3	1.1 E-1
707-1000 µm	2.3 E-1	1.9 E-1		9.2 E-2	2.8 E-1
297-707 µm	1.9 E-1	3.4 E-1		--c	3.9 E-1
149-297 µm	3.2 E-1	1.9 E-1			2.1 E-1
74-149 µm	7.3 E-1	4.7 E-2			1.0
30-74 µm	7.9 E-1	9.2 E-1			1.0
<30 µm	9.3 E-1	1.1			

a.   The ratio of zirconium to uranium in the total core is 2.47 E-1.

b.   Either zirconium or uranium was not measurable.

c.   No particle size analysis performed for smaller sizes.

core. This result indicates that aluminum is not evenly dispersed
in the core debris.

The presence of measurable quantities of gadolinium in the core
debris samples is interesting because the core inventory is only
13.5 kg. The average concentration of gadolinium in the debris samples (∿0.33 mg/g) is higher than the core inventory concentration
of 0.11 mg/g. Gadolinium (melting point ∿3000 K) is not evenly
distributed among the samples analyzed because it was measurable in
only 25% of the samples. One possible explanation for such an even
distribution of gadolinium and local accumulations is melting. An
evaluation of core relocation is being performed based on the original locations of the four gadolinium-containing assemblies and the
sampled locations.

Structural Material. The principal structural material components
measurable in the majority of core debris samples are iron, chromium,
nickel, manganese, aluminum, silicon, and tin, the major components
of stainless steel and Inconel-718. The data indicate that the
structural components are well dispersed throughout the core debris
because the individual components are measurable in most samples.
The average iron concentration is 7.6 mg iron per gram of debris, as
compared to 27 mg/g if the core inventory of iron had been evenly
distributed in the debris. These data indicate that ∿28% of the
structural steel inventory of the core has been incorporated into
the debris if the concentrations are extrapolated to the mass of the
core. Calculations were also performed for the other principal components of stainless steel and Inconel. The fractions of core inventory retained in the core debris for other structural components, as
extrapolated from the debris sample concentrations, are chromium
(52%), nickel (57%), manganese (86%), and molybdenum (97%).

Silicon is a component of stainless steel, and tin is a minor
component (1.6%) of the zircaloy fuel rods. Both elements are measurable in most samples, indicating dispersion of both the zircaloy
fuel rod cladding and the structural steel components.

Results of Radiochemical Analysis

The results of radiochemical analysis evaluated in this paper are the
gamma spectroscopy results for the bulk core samples and the stron-
tium-90 and iodine-129 analysis results for specific particles and
aliquots. Table VII lists average radionuclide concentrations for
the bulk core debris samples. With few exceptions, the concentrations are consistent within a factor of two from one sample to the
next, indicating that fission product concentrations are relatively
consistent throughout the sampling depth of the debris bed.

Table VIII compares the measured concentrations with core
inventories for each radionuclide as calculated using ORIGEN2 (5).
Table III shows that the measured core inventory of cerium-144 as
extrapolated from the measured concentrations is 1.3 times the calculated core inventory. Also, the range indicates that the higher
concentrations are consistent for all samples analyzed. Two explan-
ations for the higher concentrations of cerium-144 are possible.
The first is that the high burnup fuel material from near the center
of the core was on the surface of the debris bed . An ORIGEN2 analysis indicates that variation in burnup from the core average by a

TABLE VII. AVERAGE RADIONUCLIDE CONCENTRATIONS FOR COMPOSITE SAMPLES

Radionuclide	Average Radionuclide Concentration (μCi/g)[a]				
	Sample #1 H8-Surface (23.6 g)	Sample #3 H8-22 Inches (50.10 g)	Sample #4 E9-Surface (5.1 g)	Sample #5 E9-2 Inches (16.6 g)	Sample #6 E9-22 Inches (44.43 g)
Co-60	5.0 E+1	2.15 E+1	5.70 E+1	4.69 E+1	3.85 E+1
Sr-90	3.2 E+3	6.0 E+3	6.33 E+3	6.73 E+3	5.07 E+3
Ru-106	4.91 E+2	8.70 E+2	7.14 E+2	1.01 E+3	5.65 E+2
Sb-125	1.19 E+2	1.27 E+2	8.15 E+1	8.77 E+1	1.04 E+2
I-129[b]	5.6 E-4	3.8 E-4	2.4 E-4	5.9 E-4	4.2 E-4
Cs-134	6.72 E+1	6.00 E+1	2.44 E+1	1.10 E+2	8.85 E+1
Cs-137	1.35 E+3	1.44 E+3	4.79 E+2	2.45 E+3	2.19 E+3
Ce-144	2.88 E+3	3.00 E+3	3.60 E+3	3.06 E+3	2.43 E+3
Eu-154	5.44 E+1	5.37 E+1	5.48 E+1	4.89 E+1	3.00 E+1

a. Samples decay-corrected to March 1984.

b. I-129 analysis data obtained from a sample weighted data evaluation of individual particle and aliquot results.

TABLE VIII. PERCENTAGE OF RADIONUCLIDE RETENTION IN THE GRAB SAMPLES FROM DEBRIS BED

Radionuclide	Calculated Initial Radionuclide Inventory (μCi/g)[a]	Sample Weighted Radionuclide Concentrations (μCi/g)	Percentage of Core Inventory in Grab Sample	
			Sample Weighted Fraction Percentage	Range[b] Percentage
Sr-90	5.30 E+3	5.68 E+3	107	46[c] – 98
Ru-106	9.20 E+2	7.20 E+2	78	53 – 110
Sb-125	2.96 E+2	1.11 E+2	38	27 – 43
I-129	1.83 E-3	4.38 E-4	24	13 – 32
Cs-134	2.96 E+2	7.48 E+1	25	8.2 – 37
Cs-137	6.08 E+3	1.75 E+3	29	8 – 41
Ce-144	2.20 E+3	2.82 E+3	130	110 – 160
Eu-154	5.10 E+1	4.57 E+1	90	59 – 107

a. Core inventory on March 1, 1984, from an ORIGEN code analysis based on individual fuel rod assemblies.(5) The inventory has been extrapolated to the mass of the core (1.25 E5 kg) to provide an average concentration per gram of debris.

b. The range is calculated using the lowest and highest concentrations from Table VII.

factor of two is possible for the TMI-2 core. Another possible explanation for the high radionuclide concentrations is that cerium
vapor has condensed in the portion of the core debris sampled.
Higher volatility for some cerium oxides has been shown to be possible using thermodynamic calculations (8). It has been reported
that the rare earths (e.g., cerium) exhibit unexpectedly high vapor
pressures under conditions of high temperature and low oxygen (10).
Therefore, if this explanation is valid, some locations in the core
must be depleted in the rare earth radionuclides.

Cesium-137 and iodine-129 have similar retention percentages in
the debris bed, 29% and 24% respectively. The range of concentrations is relatively small for both radionuclides, indicating that
the weighted concentrations are probably characteristic of the upper
portion of the debris bed.

Strontium-90 and antimony-125 have retention percentages of
107% and 38%, respectively. Strontium-90 has a narrow range of retention fractions, indicating the potential for significant retention in the debris bed. Antimony is more volatile and is fairly
similar chemically to tellurium. Also, since antimony radionuclides
are the parents to the daughter tellurium radionuclides, antimony-125
may be used as an indicator of tellurium behavior.

More in-depth evaluations of the behavior of the measured radionuclides will be presented in Reference 5. Also, an evaluation of
the relationship between the measured radionuclide concentrations
and the elemental constituents will be performed.

Observations and Conclusions

The principal observations and conclusions of in this paper are as
follows:

- Samples are probably not representative of the entire core.

- The core debris samples are composites containing fuel
  materials, control rod and poison rod materials, and
  structural materials.

- The Zr/U ratios indicate that the fuel and cladding
  materials are relatively well mixed in the debris.

- Silver from the control rods is generally dispersed
  throughout the core debris samples with several localized
  accumulations.

- The extrapolated fractions of control rod materials
  retained in the core debris are silver (11%), indium
  (67%), and cadmium (49%).

- The ratio of indium to silver in the debris is higher (a
  factor of 10-40) than was present in the control rods.

- Cadmium is present in a few particles, but in a much
  higher concentration than would be found in the control
  rods.

o   Based upon the concentration of iron in the core debris
    samples, as much as 28% of the structural steel in the
    core may be incorporated in the debris bed.

o   Aluminum from the $B_4C-Al_2O_3$ rods is widely dispersed
    in the core with little or no measurable boron present,
    indicating that the material in the poison rods was
    redistributed during the accident.

o   Gadolinium (melting point ∿3000 K) appears to be un-
    evenly distributed in the debris bed, indicating the
    possibility of melting.

o   The concentration of the rare earth cerium-144 in the
    debris samples is higher than would be expected from cal-
    culated core inventories by a factor of 1.3, possibly due
    to variations in the calculated inventory or relocation of
    high burnup core material.

o   Cesium-137 and iodine-129 have similar retention fractions
    of 29% and 24%, respectively.

Literature Cited

1. Akers, D. W.; Cook, B. A. "Preliminary Report: TMI-2 Core
    Debris Grab Samples-Analysis of First Group of Samples",
    EGG-TMI-6630, June 1984.
2. Akers, D. W.; Johnson, D. A. "TMI-2 Core Debris-Cesium Release/
    Settling Test", GEND-INF-60, Volume 3, December 1984.
3. Hayner, G. O. "TMI-2 H8A Core Debris Sample Examination Final
    Report", GEND-INF-060, Vol. II, May 1985.
4. Schuman, R. P. et al. "TMI Analytical Procedures Manual", EG&G
    Idaho document, to be published.
5. Akers, D. W. et al. "TMI Core Debris Final Report",
    GEND-INF-060, Vol. 5, to be published.
6. Evans, D. L. "TMI-2 Fuel Recovery Plant Feasibility Study",
    EGG-TMI-6130, December 1982.
7. Wichner, R. P.; Spence, R. D. "Quantity and Nature of LWR
    Aerosols Produced in the Pressure Vessel During Core Heatup
    Accidents—A Chemical Equilibrium Estimate", NUREG/CR-3181,
    March 1984.
8. Coleman, D. "TMI-2 Accident Core Heat-up Analysis, A
    Supplement", NSAC 25, June 1981.
9. Vinjamuri, K. et al. "Examination of H8 and B8 Leadscrews from
    Three Mile Island Unit 2 (TMI-2)", EGG-TMI-6685, to be
    published.
10. Pearson, R. L.; Lindema, T. B. "Simulated Fission Product Oxide
    Behavior in Triso-Coated HTGR Fuel", ORNL/TMI 6741, August 1979.

RECEIVED September 26, 1985

9

Reactor Building Source Term Measurements

C. V. McIsaac and D. G. Keefer

Idaho National Engineering Laboratory, EG&G Idaho, Inc., Idaho Falls, ID 83415

Results of radiochemical, elemental, and particle size
analyses of samples collected from Three Mile Island
Unit-2 (TMI-2) are presented. The total quantities of
fission products, fuel, and core material elements
measured on Reactor Building surfaces, in the water and
sediment in the basement, and in the reactor coolant
system water during basement sampling are also presented. These measurements show that (a) 59% of the
H-3, 2.7% of the Sr-90, 0.3% of the Sb-125, 15% of the
I-129, 20% of the I-131, and 42% of the Cs-137 originally in the core at the time of the accident could be
accounted for outside the core but within the Reactor
Building, (b) iodine in solution in the water in the
basement was predominantly iodide, (c) about 60% of the
iodine in solution in the basement water in August 1979
was not in solution by March 1981, and (d) control rod
material elements were transported to the reactor coolant drain tank as hydrosols rather than as fractionated
bulk material.

This paper summarizes the total quantities of fission products, fuel,
and core structural material elements that were measured within the
TMI-2 Reactor Building, but outside the active core region. The
results of analyses of samples collected from the Reactor Building
from August 1979 through December 1983 have been evaluated and
included in the summary. The types of samples that were collected
during this time period include (a) water and sediment samples from
the basement, (b) surface samples from upper-level floors and walls,
(c) water and particulate debris samples from the reactor coolant
drain tank (RCDT), and (d) water samples from the reactor coolant
system (RCS).

0097-6156/86/0293-0168$06.00/0
© 1986 American Chemical Society

The TMI-2 plant had been operating for about 95 effective full-power days when it shut down at 0400 on March 23, 1979. Shortly after the reactor tripped, the power-operated relief valve (PORV) on the RCS pressurizer opened due to high system pressure and coolant began to escape to the Reactor Building basement. When the RCS pressure dropped to below the low-level set point of the PORV, the PORV failed to close. It was not until 0620 that the operators realized that the PORV remained open and closed it. During the time the PORV was open, sufficient coolant was lost from the RCS to uncover the core. Following core uncovery, significant amounts of hydrogen gas and fission products were released to the RCS as a result of the zircaloy fuel rod cladding oxidation and heating of the uranium dioxide fuel (1).

During the course of recovery, operators intermittently opened the PORV to control RCS pressure. During the first three days following the onset of the accident, an estimated 1.0×10^6 L of highly contaminated primary coolant escaped from the RCS to the Reactor Building basement (2). The Reactor Building atmosphere likely remained saturated with water for months following the accident, resulting in extensive pooling of contaminated water on upper-level horizontal surfaces. During these months, the volume of contaminated water in the basement continued to increase because the RCS and the river water cooling system, which was supplying water to the Reactor Building air cooling assembly, were leaking. The water in the basement eventually reached a depth of about 2.6 m by September 1981 (2).

The first sampling of the water and sediment in the Reactor Building basement took place in August 1979 when the volume of water in the basement was about 2×10^6 L. Since that time, liquid and sediment samples have been collected from the basement at six other locations; the most current samples were collected in August 1983. In addition, similar samples were collected from the RCDT in December 1983. Several regions of the basement, including the inside of the RCDT, were also visually examined in 1983 using a closed-circuit television system.

The first systematic samplings of Reactor Building surfaces took place in December 1981 and March 1982, before and after the Reactor Building gross decontamination experiment. One hundred eighty samples were collected from Reactor Building structural surfaces such as floors and D-ring walls. Using a different sampling technique, additional surface samples were collected in September 1983 to determine the depth activity had penetrated into the painted structural concrete. Seventeen core samples were removed from Reactor Building concrete floors and D-ring walls.

In October 1981, in-situ gamma spectral measurements were made of the total activities deposited on the surfaces of three of the five cooling coils that are installed in the Reactor Building air cooling assembly. To augment these surface activity data, a number of scrapes and smears were collected in 1983 from various external and internal surfaces of the assembly. In addition, the five cooling coil access panels were removed from the assembly for subsequent laboratory analysis.

Flooding of the Reactor Building Basement

The three floors in the Reactor Building are designated by their
elevations, in feet, above sea level. Thus, the basement floor,
which is at an elevation of 282 ft, 6 in., is commonly known as the
282-ft floor. The basement is the area between the 282-ft and
305-ft floors. A number of cubicles located in the basement shield
components such as the RCDT and pump, leakage coolers, leakage
transfer pumps, letdown coolers, and sump. All of these components
except the sump are at or above the level of the basement floor.
The volume of the RCDT, 2.74×10^4 L, is about three times the
volume of the Reactor Building sump.
     When the first batch of water was pumped out of the basement
on September 23, 1981, for processing through the Submerged Demineralizer System (SDS) and EPICOR II ion-exchange resins, the water
level in the basement had reached about 2.6 m. This water is
attributed to three major sources: the RCS, the Reactor Building
spray system, and the river water cooling system.

Accident Water. The TMI-2 plant had been operating for about three
months when the series of events that led to the accident began at
0400 on March 28, 1979. At about that time, the pumps that normally
supply feedwater to the steam generators tripped, thereby stopping
normal feedwater flow. The auxiliary feedwater pumps automatically
sequenced on when the main pumps tripped, but because the block
valves downstream of the auxiliary pumps were closed, emergency
feedwater flow to the steam generators was not initiated. When the
flow of feedwater was interrupted, the level of secondary water in
the steam generators dropped rapidly, thereby reducing their capacity to remove heat from the RCS. As a result, the pressure in the
RCS increased dramatically and the reactor automatically shut down.
     The pressure in the RCS quickly increased beyond the high-level
set point of the PORV, which is located on the pressurizer. When
this relief valve opened, reactor coolant began to escape from the
RCS. The lost coolant flowed through a 36-cm-diameter pipe to the
RCDT located in the basement some 18 m below the level of the PORV.
As a consequence of rapid pressurization, the rupture disk on this
tank burst and coolant escaped to the basement floor through a
46-cm-diameter pipe that encloses the rupture disk. The rupture
line rises vertically from the top of the tank, turns 90 degrees,
and terminates in a penetration in the west wall of the RCDT cubicle. Coolant continued to escape to the Reactor Building basement
via this pathway until 0620 when the PORV block valve was closed.
Additional coolant, in the form of steam and water, and hydrogen gas
escaped through the PORV from 0713 to 1700 hours, when the block
valve was intermittently opened to regulate RCS pressure. An estimated 1×10^6 L of reactor coolant was released to the basement
via this same pathway during the first three days following the
onset of the accident (2).
     In addition to the 1×10^6 L of RCS water released during the
accident, an average of 29.5 L/h flowed through the PORV block valve
for more than two years following the accident. This leakage contributed 6.74×10^5 L of RCS water to the basement water volume (2). Thus, the total volume of RCS water that escaped to the

basement was approximately 1.67 x 10^6 L, which is about 69% of the
total volume of water released to the basement as of September 23,
1981.
   As a result of the hydrogen burn pressure spike that occurred
at 1350 hours on the day of the accident, the Reactor Building spray
system activated and remained on for approximately 6 min. During
that time, the system discharged an estimated 6.43 x 10^4 L of
chemically treated water, containing boron and sodium hydroxide,
into the Reactor Building atmosphere (2). Sodium hydroxide is added
to the water to remove halogens (i.e., I and Br). Upper-level
radiation monitors registered decreases following the spray,
indicating that the sprayed water effectively removed at least some
of the airborne contaminants. Most of this water probably eventually drained to the basement. The volume of water discharged by
the spray system represents about 3% of the total basement water
volume as of September 23, 1981.
   Further increases in the basement water level after the accident are attributed to leakage from the river water cooling system
of the Reactor Building air cooling assembly. The leakage is
suspected to have been from a relief valve on the assembly cooling
coils. Based on back projections of water level and reconstruction
of events associated with water inventory, an estimated 6.81 x 10^5 L
of river water was released to the basement from this source before
it was secured (2). The river water from this source represents
about 28% of the maximum basement water inventory prior to the start
of SDS processing in September 1981.

Decontamination Water. After most of the initial accident water was
removed from the basement and processed through the SDS and EPICOR II
systems, some of the water that was processed was staged for use in
decontaminating the Reactor Building. This water, in its undiluted
state, began to be recycled to the Reactor Building in March 1982
when upper-level floors and walls, cable trays, and major pieces of
equipment were sprayed with high- and low-pressure processed water.
Most of this decontamination water drained to the basement, carrying
with it the fission products that were washed from the structural
and equipment surfaces.
   The depth of the water in the Reactor Building basement from
May 1979 through December 1983 is shown graphically in Figure 1.
Prior to the start of SDS processing on September 23, 1981, the water
level had been increasing at a fairly constant rate due to leakage
from the RCS and the river water cooling system. However, by the
time the gross decontamination experiment commenced six months later
in March 1982, about 2.3 x 10^6 L of contaminated water had been
pumped from the basement and processed through the SDS. This initial
processing, which was done in 16 separate batches, reduced the water
depth to about 17 cm. The gross decontamination experiment and subsequent decontamination operations periodically increased the water
depth. By mid-April 1983, an estimated 1.4 x 10^6 L of processed
water had been used for decontamination purposes and had returned to
the basement.

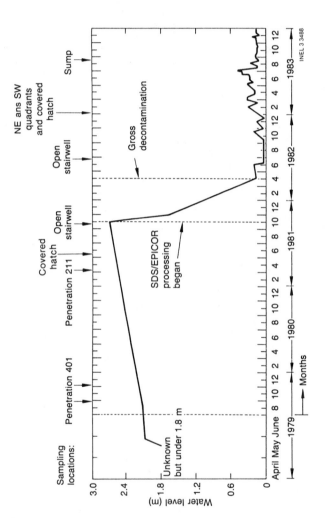

Figure 1. Reactor Building water level during the period the basement was sampled.

Reactor Building Basement Samples

The volumes and masses of the liquid and solid fractions of the
basement samples that have been analyzed are summarized in Table I.
To date, 24 samples have been acquired using a variety of sampling
techniques. Samples obtained prior to the start of SDS processing
were collected to determine the unperturbed inventories of fission
products and core materials in the water and sediment in the basement. Because SDS processing removed the majority of the cesium and
strontium activity from the water that was used for Reactor Building
decontamination, samplings performed since March 1982 focused on
characterizing the sediment present on the basement floor. The
locations where liquid and sediment samples have been collected to
date are shown in Figure 2, which is the floor plan of the Reactor
Building basement.

Reactor Building Surface Samples

The total area of the exposed surfaces within the Reactor Building
is estimated to be about 2.2×10^4 m^2 ($\underline{3}$). The majority of the
surfaces are protected by coatings of expoxy-based, nuclear grade
paints. Surface samples were obtained from the floors and walls at
the 305-, 347-, and 367-ft elevations and from the air cooling
assembly.

Structural Surfaces. To measure the effectiveness of the Reactor
Building gross decontamination experiment performed in March 1982,
85 surface samples were collected from Reactor Building structural
surfaces in December 1981, and an additional 95 surface samples were
obtained from the same surfaces in late March 1982 following the
completion of the decontamination experiment ($\underline{3}$). The samples were
collected using a milling tool that was designed to allow sampling
over a range of depths. Paint shavings and concrete dust or metal
shavings generated during milling were, in each case, swept from the
surface being sampled and collected on a sample collection filter.

Air Cooling Assembly Surfaces. To understand better the fission
product transport mechanisms that were in effect in the Reactor
Building atmosphere, surface samples were obtained from the five
Reactor Building air coolers. A variety of samples were collected
and measurements made, including particulate scrapes and smears from
various locations, metal coupons removed from the access panels, and
in-situ gamma scans of the cooling coils and drip pans.

Discussion of Analyses Results

The concentrations of radionuclides and stable elements in the water
and particulate debris in the basement and on the surfaces of the air
cooling assembly are discussed in this section. These results are
combined with similar results for RCS water and Reactor Building
structural surfaces, and estimates are made of the total quantities
of fission products and core materials that were retained within the
Reactor Building.

THE THREE MILE ISLAND ACCIDENT

Table I. Volumes and Masses of Basement Samples That Have
Been Analyzed

Before SDS Processing			After SDS Processing		
Sampling Date (m/d/y)	Sample Volume (mL)	Mass of Filtered Solids (mg)	Sampling Date (m/d/y)	Sample Volume (mL)	Mass of Filtered Solids (mg)
8/28/79	30	--[a]	6/23/82	18	389.4
8/28/79	30	--[a]	6/23/82	27.5	718.6
8/28/79	30	76.2[b]	1/11/83	45	2
11/15/79	1050	333[b]	--	--	--
3/19/81	1000	--[a]	--	--	--
3/19/81	1000	--[a]	1/11/83	55	<1
3/19/81	1000	--[a]	1/11/83	55	497
5/14/81	85	--[a]	8/22/83	200	71.8
5/14/81	105	--[a]	12/05/83	120	--[a]
5/14/81	110	--[a]	12/12/83	128	9.1
5/14/81	110	108	--	--	--
9/24/81	120	24.6	--	--	--

a. Not measured. Insufficient solids.

b. Mass was estimated based on volume of centrifuged solids.

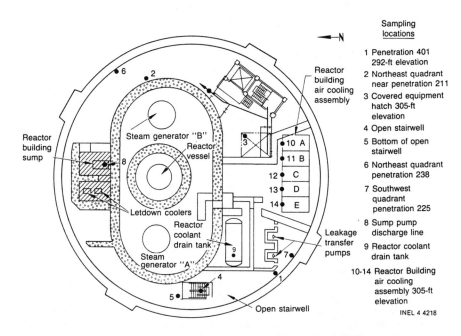

Figure 2.   Reactor Building basement and air cooling access panel
sample locations.

Basement Samples. The radiochemical, gamma-ray spectroscopy, and elemental analyses results for the samples collected from the Reactor Building basement from August 1979 through August 1983 are summarized in Tables II and III, along with similar results for the RCDT samples. The results for the liquid portions of the samples are given in Table II, and those for the solids portions are given in Table III. The radionuclide concentrations presented in the tables are decay-corrected to the sample collection dates.

The analyses results for the three water samples collected on March 19, 1981, from beneath penetration 211 are not presented in Table II because only a few of the results were available at the time of this writing. The concentrations of Cs-134, Cs-137, and I-129 were reported as 19.2 ± 0.2, 138 ± 3, and 5.5×10^{-6} μCi/mL, respectively, decay-corrected to the day the sampling took place ($\underline{4},\underline{5}$). The concentrations of these three radionuclides in the water beneath penetration 211 are essentially identical to their respective concentrations in the samples collected May 14, 1981, from beneath the covered hatch. Since these two sampling locations are on opposite sides of D-ring B, the good agreement suggests that soluble radionuclides were homogeneously dispersed in the water throughout large regions of the basement.

By the time the basement water was first sampled on August 28, 1979, the concentrations of H-3, Cs-134, and Cs-137 in the RCS water that was leaking to the basement had decreased to values substantially lower than their respective concentrations in the basement water ($\underline{6}$). On the other hand, the concentration of Sr-90 in the RCS water being added to the basement remained at least a factor of two higher than its concentration in the basement water throughout the 1979 to 1981 sampling period ($\underline{6}$). As a result, the effect of RCS leakage during this time was to decrease the concentrations of H-3, Cs-134, and Cs-137 and increase the concentration of Sr-90 in the basement water. The concentrations of H-3 and Cs-137 in the basement water decreased from initial values of 1 and 176 μCi/mL, respectively, on August 28, 1979, to about 0.6 and 137 μCi/mL on September 24, 1981. After being corrected for radioactive decay, the concentrations of H-3 and Cs-137 decreased by 36 and 18%, respectively. During this same time period, the concentration of Sr-90 increased from 2.8 to about 5 μCi/mL.

The concentration of boron in the RCS coolant fluctuated between 2900 and 4300 μg/mL during this same time period, and its time-weighted average concentration was about 3600 μg/mL ($\underline{6}$), which is approximately a factor of two higher than its mean concentration in the basement water. The data presented in Table II show that the concentrations of B and Na in the basement water remained essentially constant throughout the 19 months prior to the start of basement water processing through the SDS. Their mean concentrations in the basement water during this period were 2100 μg/mL ± 7% and 1200 μg/mL ± 2.5%, respectively, where the uncertainties are standard deviations at the one-sigma level. Since the river water that leaked from the air cooling assembly entered the basement on the east side of the building and contained only trace quantities of boron and the RCS coolant entered on the west side, outside the RCDT cubicle, the constancy of the concentrations of B and Na at several different sampling locations is additional evidence that soluble elements and compounds were uniformly dispersed in the basement water.

It is evident from an examination of the data presented in Table II that the concentrations of dissolved I-129 measured in the samples collected from beneath penetration 211 on March 19, 1981, and from beneath the covered hatch on May 14, 1981, are factors of three to five lower than the concentrations of dissolved I-129 measured in the water samples collected through penetration 401 on August 28, 1979, and from the floor beneath the open stairwell on September 24, 1981. The concentrations of dissolved I-129 at these four locations were measured to be 5.5×10^{-6}, 4.3×10^{-6}, 1.38×10^{-5}, and 2×10^{-5} µCi/mL, respectively. It should be noted that the concentrations given for the first three locations are, in each case, averages of the concentrations measured in three or more samples, whereas the concentration cited for the fourth location, the open stairwell, is the result of a measurement made on a single sample collected from the elevation of the floor. An investigation of the analysis method used on the sample collected September 24, 1981, from the bottom of the open stairwell determined that the liquid was not filtered prior to analysis for I-129(7). Based on the concentration of I-129 in the solids fractions of the open stairwell samples, only a very modest quantity of solids in the sample would have seriously biased the soluble I-129 result. For this reason, the concentration of 2×10^{-5} µCi/mL for the September 24 sample must be rejected. If we ignore the result for the open stairwell, the data indicate that the concentration of dissolved I-129 decreased from 1.38×10^{-5} µCi/mL in August 1979 to about 5×10^{-6} µCi/mL by March 1981.

The results of a chemical species analysis that was performed on the three 1-L samples of water that were drawn from the basement through penetration 211 in March 1981 indicate that the iodine in solution was predominantly iodide (5). The results of that analysis are as follows:

Iodine Species	I-129 Concentration (µg/mL)
Total iodine	0.031
Iodide	0.030
Iodate	0.00056
Elemental	0.00061

If radioiodine was released from the core and entered the water as elemental iodine (I_2), then iodide (I^-) would have immediately accounted for about 50% of the total iodine in solution, according to the reaction

$$I_2 + H_2O = I^- + H^+ + HOI \tag{1}$$

The fraction that was iodide would have increased to about 83% after a few days had elapsed, according to the overall reaction

$$3I_2 + 3H_2O = 5I^- + 6H^+ + IO_3^- \tag{2}$$

Iodate (IO_3^-) would not have been present initially, but after a few days, it would have accounted for about 17% of the iodine

Table II. Summary of Analyses Results for
(Activities Decay-corrected

Sampling location:	Penetration 401		Covered Hatch	Open Stairwell
Sample collection date:	8/28/79	11/15/79	5/14/81	9/24/81
Basement water volume (mL):	1.95 E+9	2.01 E+9	2.36 E+9	2.38 E+9
Nuclide concentrations (μCi/mL):				
H-3	1.03 E+0	1.04 E+0	6.03 E-1	5.87 E-1
Sr-90	2.81 E+0	2.3 E+0	5.2 E+0	4.8 E+0
Ru-106	7.0 E-3	3.1 E-3	--d	--d
Sb-125	1.5 E-2	2.3 E-2	3.0 E-2	<2 E-2
I-129	1.38 E-5	--c	4.3 E-6	2 E-5
Cs-134	3.99 E+1	3.21 E+1	1.92 E+1	1.62 E+1
Cs-137	1.76 E+2	1.62 E+2	1.43 E+2	1.37 E+2
Ce-144	6.3 E-3	8.7 E-4	--d	<3.5 E-2
Fuel concentrations (μg/mL):				
U	1.6 E-2	--c	--c	<3 E-2
Pu	1.8 E-5	--c	2.2 E-4	--c
Element concentrations (μg/mL):				
Li	1.53 E+0	--c	1.8 E+0	1.8 E+0
B	2.02 E+3	2.0 E+3	2.12 E+3	2.30 E+3
Na	1.16 E+3	1.2 E+3	1.18 E+3	1.24 E+3
Mg	<2 E+0	--c	5.4 E+0	7.3 E+0
Al	3 E+0	--c	1.4 E+0	1.8 E+0
Si	--c	--c	4.8 E+0	6.8 E+0
K	4 E+0	--c	1.6 E+1	2.0 E+1
Ca	9.3 E+0	--c	3.6 E+1	4.1 E+1
Cr	7 E-1	--c	2 E-1	3 E-1
Mn	<1 E-1	--c	<2 E-1	<2 E-1
Fe	1.2 E+0	--c	9 E-1	7 E-1
Co	<1 E-1	--c	<5 E+0	<5 E+0
Ni	<1 E+0	--c	<1 E+0	1.1 E+0
Cu	<3 E+0	--c	<1 E+0	<1 E+0
Zn	4.7 E-1	--c	<5 E+0	<5 E+0
Zr	--c	--c	<1 E+0	1.4 E+0
Ag	<3 E-1	--c	<1 E+0	<1 E+0
Cd	<2 E-1	--c	<2 E+0	<2 E+0
In	<1 E-1	--c	<5 E+0	<5 E+0
Sn	--c	--c	<5 E+0	<5 E+0
Gd	--c	--c	<1 E+0	<1 E+0

a. Volume of reactor building sump
b. Volume of reactor coolant drain tank
c. Not measured.
d. Not detected.

quid Fractions of Reactor Building Basement Samples
Sample Collection Dates)

Open Stairwell		Covered Hatch	Penetration 238	Penetration 225	Sump	Reactor Coolant Drain Tank
23/82	6/23/82	1/11/83	1/11/83	1/11/83	8/22/83	12/12/83
99 E+7	9.99 E+7	1.83 E+8	1.83 E+8	1.83 E+8	(1.03 E+7)[a]	(2.74 E+7)[b]
--[c]	--[c]	--[c]	--[c]	--[c]	--[c]	3.5 E-2
98 E+0	5.8 E+0	2.49 E+0	2.36 E+0	3.45 E+0	6.1 E+0	2.51 E+0
.6 E-2	2.6 E-3	<1.4 E-2	<3.7 E-2	<9.4 E-3	--[d]	<2.7 E-2
<3 E-3	2.6 E-2	<7 E-3	<2 E-2	<5 E-3	--[d]	6.4 E-3
<2 E-4	--[c]	1.08 E-6	1.11 E-6	4.09 E-7	--[d]	3.9 E-7
40 E+1	1.44 E+1	9.8 E-1	6.7 E-1	6.6 E-1	6.16 E+0	7.73 E-2
51 E+2	1.59 E+2	1.20 E+1	8.22 E+0	8.26 E+0	9.55 E+1	1.23 E+0
--[d]	--[d]	<4.2 E-3	<1.2 E-2	<3.0 E-3	--[d]	<1.5 E-2
.6 E-2	5 E-3	--[c]	5.7 E-2	--[c]	5.7 E-2	<1.3 E-1
.6 E-5	<1 E-4	--[c]	5.3 E-8	--[c]	5.3 E-8	--[c]
--[c]	1.0 E+1	--[c]	--[c]	--[c]	2.35 E+0	--[c]
∿3 E+3	8.0 E+3	--[c]	--[c]	--[c]	7.97 E+3	--[c]
∿3 E+3	6.0 E+3	--[c]	--[c]	--[c]	2.56 E+3	--[c]
5 E+0	8.0 E+1	--[c]	--[c]	--[c]	6.80 E+0	--[c]
3 E+0	1.7 E+0	--[c]	--[c]	--[c]	3.16 E+0	--[c]
,0 E+1	9.0 E+1	--[c]	--[c]	--[c]	8.22 E+0	--[c]
,0 E+1	2.0 E+2	--[c]	--[c]	--[c]	2.50 E+1	--[c]
,0 E+1	2.0 E+1	--[c]	--[c]	--[c]	2.66 E+1	--[c]
3 E+0	2 E+0	--[c]	--[c]	--[c]	2.50 E-2	--[c]
2 E-1	--[d]	--[c]	--[c]	--[c]	1.42 E+0	--[c]
6 E-1	3 E+0	--[c]	--[c]	--[c]	1.36 E+0	--[c]
1 E-1	--[d]	--[c]	--[c]	--[c]	2.90 E-2	--[c]
5 E-1	--[d]	--[c]	--[c]	--[c]	6.63 E-1	--[c]
5 E+0	6 E-1	--[c]	--[c]	--[c]	9.59 E+0	--[c]
,-[c]	--[d]	--[c]	--[c]	--[c]	5.57 E-1	--[c]
<4 E-1	--[d]	--[c]	--[c]	--[c]	<5 E-2	--[c]
1 E-1	4 E-2	--[c]	--[c]	--[c]	1.30 E-1	--[c]
2 E+0	1.7 E-1	--[c]	--[c]	--[c]	2.08 E-1	--[c]
1 E+0	8 E-2	--[c]	--[c]	--[c]	<5 E-1	--[c]
<2 E+0	2 E-2	--[c]	--[c]	--[c]	3.46 E-1	--[c]
,-[c]	--[d]	--[c]	--[c]	--[c]	<1 E-1	--[c]

Table III. Summary of Analyses Results for
(Activities Decay-corrected

Sampling location:	Penetration 401		Covered Hatch	Open Stairwell
Sample collection date:	8/28/79	11/15/79	5/14/81	9/24/81
Nuclide concentrations (μCi/g):				
Co-60	5.2 E-1	1.3 E+0	1.21 E+1	2 E+1
Sr-90	8.2 E+1	6.1 E+2	8 E+2	2.2 E+3
Ru-106	1.7 E+1	1.8 E+1	1.04 E+2	5.8 E+1
Sb-125	1.1 E+1	2.8 E+1	4.87 E+2	1.2 E+1
I-129	4.8 E-3	--b	1.1 E-1	2.7 E-3
Cs-134	5.7 E+0	1.5 E+1	1.07 E+2	3.9 E+1
Cs-137	2.5 E+1	7.6 E+1	8.08 E+2	3.24 E+2
Ce-144	1.2 E+1	3.9 E+1	6.6 E+1	9.4 E+1
Fuel concentrations:				
U (mg/g)	4.3 E-2	3.2 E-1	4.0 E+0	3.9 E-1
Pu (μg/g)	6.3 E-2	--b	2.9 E+0	--b
U-235 (at.%)	2.35 E+0	--b	2.6 E+0	<4 E+0
Element concentrations (μg/g):				
Li	<1.2 E+2	--a	--a	--a
B	1.2 E+3	--b	4 E+4	1.4 E+5
Na	<4 E+2	--a	1.9 E+3	--b
Mg	2.7 E+3	2.3 E+3	2 E+3	4 E+3
Al	3.2 E+3	2.3 E+4	1.1 E+4	5 E+4
Si	--b	1.1 E+4	7 E+4	3 E+4
K	4.0 E+2	2.3 E+2	1.7 E+3	--b
Ca	7.8 E+2	7.1 E+3	2 E+4	4 E+4
Cr	7.8 E+2	4.0 E+2	1 E+4	3 E+4
Mn	4.0 E+2	1.5 E+3	2.4 E+3	2 E+4
Fe	4.0 E+3	1.4 E+4	1.5 E+5	1.2 E+5
Co	<4 E+1	5.5 E+1	--a	--a
Ni	4.0 E+3	4.0 E+4	3 E+4	2.5 E+4
Cu	2.2 E+4	1.2 E+5	1.0 E+5	3 E+3
Zn	7.8 E+2	6.3 E+3	1.8 E+4	--b
Zr	--a	1.1 E+2	--a	--a
Ag	3.2 E+3	8.6 E+2	1.6 E+4	--b
Cd	<2 E+2	8.6 E+2	6.4 E+3	--b
In	1.2 E+2	4.7 E+2	1 E+4	--b
Sn	--a	6.3 E+2	--a	--a
Gd	--a	--a	--a	--a
Total weight percent:	4.6	23.1	49.3	46.2

a. Not detected.
b. Not measured.

Solids Fractions of Reactor Building Basement Samples
to Sample Collection Dates)

Open Stairwell	Open Stairwell	Covered Hatch	Penetration 238	Penetration 225	Sump	Reactor Coolant Drain Tank
6/23/82	6/23/82	1/11/83	1/11/83	1/11/83	8/22/83	12/12/83
9.6 E+0	1.17 E+1	<8.0 E-1	>3.3 E+1	1.47 E+0		
2.38 E+3	4.9 E+3	1.2 E+3	>5.7 E+2	8.38 E+1	2.48 E+0	3.19 E+1
--a	4.1 E+1	<1.2 E+1	>5.9 E+1	2.72 E+0	1.51 E+2	1.40 E+4
1.33 E+2	1.42 E+2	1.7 E+1	>3.6 E+2	1.13 E+1	--a	7.8 E+1
1.8 E-1	--b	1.6 E-2	--b	1.97 E-3	1.7 E+0	1.72 E+1
7.43 E+1	1.84 E+2	3.2 E+2	>4.2 E+1	2.30 E+0	<2.5 E-5	5.2 E-8
8.02 E+2	2.04 E+3	3.6 E+3	>5.4 E+2	2.99 E+1	3.55 E+0	6.4 E+0
--a	5.2 E+1	2.6 E+1	>1.3 E+1	2.03 E+0	5.37 E+1	9.8 E+1
					1.4 E+1	1.3 E+2
2.97 E+0	3.9 E+0	2.2 E+0	--b	3.0 E+0		
4.41 E+0	6.1 E+0	3.8 E+0	>4.5 E-1	5.2 E-1	1.8 E-1	3.7 E+0
2.37 E+0	2.4 E+0	2.74 E+0	--b	--b	1.3 E+0	--b
					--b	--b
--a	5 E+0	--b	--b	--b		
∿3 E+3	1 E+4	--b	--b	--b	1.09 E+1	1.58 E+2
>3 E+3	3 E+2	--b	--b	--b	3.22 E+4	9.01 E+4
>2 E+3	5 E+3	--b	--b	--b	2.95 E+2	1.34 E+4
>5 E+3	1.2 E+4	--b	--b	--b	9.92 E+1	8.48 E+3
>2 E+4	7 E+3	--b	--b	--b	1.11 E+4	1.88 E+4
8 E+2	5 E+1	--b	--b	--b	4.86 E+2	2.60 E+3
2 E+3	3 E+3	--b	--b	--b	1.00 E+4	1.06 E+5
4 E+2	1 E+2	<6.7 E+2	<1.5 E+2	<4.4 E+2	2.30 E+2	8.17 E+4
1 E+3	5 E+1	1.6 E+3	<1 E+2	1.4 E+3	7.58 E+1	2.72 E+3
>8 E+3	3 E+3	5.4 E+3	3.9 E+3	4.8 E+3	6.25 E+1	6.40 E+3
1 E+1	2 E+1	--b	--b	--b	9.88 E+3	1.14 E+5
>1.2 E+3	8 E+3	<2 E+2	<4 E+1	2.2 E+3	6.42 E+0	2.62 E+2
>4 E+4	4.9 E+4	1.7 E+3	3.1 E+2	1.3 E+4	1.51 E+2	1.11 E+5
>2 E+3	2 E+2	3.5 E+2	<2 E+3	<6 E+2	1.42 E+2	3.22 E+5
>2 E+3	2 E+2	1.2 E+2	6.5 E+1	5.6 E+2	1.21 E+2	2.47 E+4
>5 E+3	2.5 E+4	--b	--b	--b	2.03 E+2	1.89 E+3
>5 E+3	5 E+3	--b	--b	--b	2.94 E+2	6.29 E+3
1.5 E+3	3 E+3	--b	--b	--b	1.47 E+2	1.19 E+4
1 E+3	1.4 E+3	--b	--b	--b	<7 E+1	6.20 E+3
--1	--1	--b	--b	--b	<4 E+1	4.96 E+3
					<1 E+1	1.95 E+2
10.6	13.6	1.0	0.7	2.6		
					6.6	93.8

in solution (5). The measured concentration of iodate corresponds
to 1.8% of the total iodine in solution and is therefore a factor of
10 less than the fraction expected if iodine entered the coolant as
elemental iodine. These data are not inconsistent with the conclusion that fission product iodine entered the water as iodide and not
as elemental iodine.

The data presented in Table III for the solids fractions of the
Reactor Building basement samples indicate that Sr-90 is the predominant activity in both the sediment on the basement floor and the
particulate matter retained in the RCDT. The average concentration
of Sr-90 in the sediment covering the basement floor is about
1.4 mCi/g (mCi per dry gram), and its concentration in the solids in
the RCDT is 14 mCi/g. The nuclide having the second highest concentration in the basement sediment is Cs-137; its average concentration
is about 0.86 mCi/g. The average concentrations of Sb-125, Cs-134,
Ru-106, and Ce-144 in the basement sediment were all about a factor
of 10 to 20 lower than the average concentration of Cs-137.

A number of compounds and elements measured to be present in the
water and sediment in the basement can combine with strontium to form
slightly soluble or insoluble compounds. An abbreviated list of
candidate strontium compounds is provided in Table IV. Boron, carbon, and silicon were detected in many of the sediment samples; and
dissolved carbonate, oxalate, and sulfate were measured in several
samples of water. Phosphate, on the other hand, was not detected in
these same samples that were analyzed using ion chromatography.

Among the elements that exhibited the highest concentrations in
the sediment, copper and silver are of special interest because they
both can combine with iodine to form insoluble or very slightly soluble compounds. Copper combines with iodide (I^-) to form CuI (or
Cu_2I_2), which has a solubility in cold water of 8 μg/mL, and
with iodate (IO_3^-) to form $Cu(OH)IO_2$ and Cu_2HIO_6, both of
which are insoluble in cold and hot water. Silver combines with
iodine to form AgI, which is insoluble, and $AgIO_3$, which has a
solubility of 30 μg/mL in cold water and 190 μg/mL in hot water.
The data presented in Table III indicate that there is a good correlation between the concentrations of Cu and Ag in the sediment on the
basement floor and the concentration of I-129 in the sediment. Sampling locations having the highest concentrations of Cu and Ag are
also the locations with the highest concentrations of I-129.

The presence of silver in the sediment on the basement floor and
in the debris in the RCDT is also of interest because silver is the
major constituent of the Ag-In-Cd alloy, which is the neutron
absorber in the majority of the reactor control rods. By weight,
the alloy is 80% Ag, 15% In, and 5% Cd. The concentration of Ag in
the sediment ranges from 2.5 wt% at the location of the open stairwell to 0.029 wt% in the sump. Its concentration in the RCDT solids
is 0.63 wt%. The other elements of the alloy, In and Cd, were also
detected in the samples collected from the basement. The concentration of In in the sediment on the basement floor varied between less
than 0.007 wt% in the sump to 0.3 wt% at the location beneath the
open stairwell. The concentration of Cd in the sediment ranged from
a low of 0.015 wt% in the sump to a high of 0.64 wt% beneath the
covered hatch. The concentrations of In and Cd in the particulate
matter collected from the RCDT were 0.62 and 1.19 wt%, respectively.

The masses of Ag, In, and Cd in the reactor core prior to the accident were 2199, 412, and 137 kg, respectively (8). On an atom basis, these masses correspond to 6 atoms of Cd and about 18 atoms of In for every 100 atoms of Ag in the core. The relative number of atoms of these three elements in the samples of solids collected from the RCDT and basement floor is given below.

Location	Ag	In	Cd
Reactor Core	100	18	6
Reactor coolant drain tank	100	93	182
Basement floor	100	82	9
Sump	100	<22	48

The fact that the relative abundances of these three elements in the RCDT and basement are different from their known values in the core is not surprising given the low melting temperature of the Ag-In-Cd alloy and the different boiling temperatures of the individual elements. The alloy melts at 799°C; and the control rod cladding, which is 304 stainless steel, melts at a temperature between 1399 and 1454°C (8). Pure Cd boils at the comparatively low temperature of 765°C at atmospheric pressure, 34°C below the melting temperature of the alloy. When core temperatures reached the melting temperature of the alloy, some Cd gas would have begun to evolve from the alloy that melted. As core temperatures escalated, Cd vapor would have accumulated in the upper plenums of the control rods. The Cd vapor pressure within the rods would have continued to increase until the cladding began to melt. Calculations that have been performed (9) indicate that the maximum vapor pressure of Cd inside the control rods was not sufficient to cause cladding failure. It has been estimated that Cd vapor, at a maximum, accounted for about 35% of the vapor pressure within the rods, and He, which is the gas used to pressurize the rods, accounted for the remaining 65% (9).

Based on analyses of core debris samples, temperatures of the fuel rods rose to peak values of between 2630 and 2830°C before high-pressure injection was resumed (10). When the control rods began to fail, presumably between 1399 and 1454°C, Cd vapor and liquefied alloy would have escaped from the rods. The boiling temperatures of Ag and In at atmospheric pressure are 2212 and 2000°C (8), respectively, but the high pressure in the RCS would have suppressed the boiling of the two elements when temperatures in the core reached these levels. However, the approximate one-to-one proportionality in the number of atoms of Ag and In measured in the sample of solids removed from the RCDT indicates that Ag and In were not transported there as particles of resolidified alloy. The data indicate that they were transported from the core as condensed vapors. The quantity of Cd in the RCDT solids is disproportionately large compared to the quantities of Ag and In found in the solids because a significantly larger fraction of the Cd in the core was vaporized compared to the fractions of the core inventories of Ag and In that were vaporized.

Cubiccotti and Sehgal (11) recently reported the results of thermodynamic calculations of the volatilities of core materials during a postulated severe light water reactor accident. Their calculations indicate that the volatilities of core materials decrease roughly in the following order: Cd, In, Sn, Fe, Ag, Mn, Ni, Cr, U, and Zr. The quantities of these elements measured in the sample of particulates that was collected from the RCDT are given below, expressed as percent of core inventory per gram of sample. The elements that make up stainless steel (i.e., Fe, Mn, Ni, and Cr) are omitted below because access to the inside of the tank was gained by boring a hole through the stainless steel pipe that encloses the tank's rupture disk. Fines created during the boring fell into the tank and likely contaminated the sample of sediment that was collected.

Element	% per gram
Cd	8.8 E-6
In	1.5 E-6
Sn	1.5 E-6
Ag	2.8 E-7
Zr	8.5 E-9
U	4.6 E-9

The concentrations listed above are in good agreement with the relative concentrations expected, based on the calculated relative volatilities of the elements.

The particle size distributions of the solids collected from the RCDT and sump were measured by analyzing 500X photomicrographs of the material filtered from the samples following ultrasonic treatment. The results for these two samples are shown as bar graphs in Figures 3 and 4, respectively. Measurements were made in 2-μm intervals, and they did not include particles under 1 μm. An obvious difference between the particle size distributions shown in the figures is the higher population of large particles in the sump. By number, about 12% of the particles collected from the sump are larger than 10 μm, but only about 6% of the particles collected from the RCDT are larger than 10 μm. Only about 39% of the sump particles were in the 1- to 3-μm size range, whereas about 63% of the RCDT particles had sizes in this range. On a population basis, the mean particle sizes measured in the sump and RCDT samples were 6.2 and 3.9 μm, respectively. The particle size distribution analysis of the sediment collected from the floor near the bottom of the open stairwell was performed using a HIAC Particle Size Analyzer. The particle size distribution measured in this sample is very similar to the one determined for the RCDT sample. Like the RCDT sample, the sediment from beneath the open stairwell was found to contain very few particles larger than 10 μm. The mean particle size on a volume basis was measured to be about 20 μm, and on a population basis it was about 4 μm, which is essentially identical to the mean particle size of the RCDT solids. Sixty-five percent of particles in this same sediment sample exhibited sizes between 1.5 and 3 μm, and only 1.5% were found to be larger than 10 μm. These particle size distributions are typical of those of aerosols. This is additional evidence that core materials were transported from the core as aerosols and/or hydrosols.

Surface Samples. Table V compares the average surface activities measured on the air cooling assembly access panels with the surface

Table IV.   Slightly Soluble or Insoluble Strontium Compounds

Compound Name	Formula	Cold Water (g/100 mL)	Hot Water (g/100 mL)
Strontium hexaboride	SrB_6	Insoluble	Insoluble
Strontium carbonate	$SrCO_3$	0.0011	0.065
Strontium hyponitrite	$SrN_2O_2 \cdot 5H_2O$	Very slightly soluble	Slightly soluble
Strontium oxalate	$SrC_2O_4 \cdot H_2O$	0.0051	5
Strontium diorthophosphate	$SrHPO_4$	Insoluble	Insoluble
Strontium metasilicate	$SrSiO_3$	Insoluble	Insoluble
Strontium monosulfide	SrS	Insoluble	Decomposes
Strontium sulfite	$SrSO_3$	0.0033	--

Source: Reproduced with permission from Ref. 13.   Copyright 1972, CRC Press.

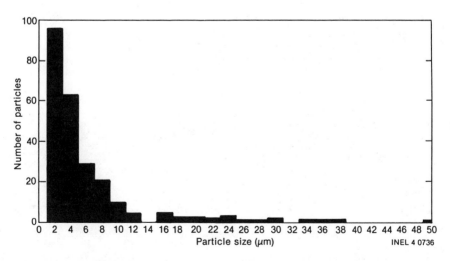

Figure 3.   Particle size distribution of sediment collected from the Reactor Building sump.

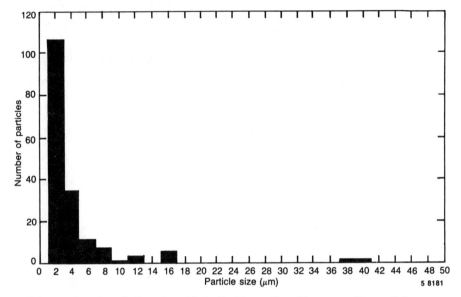

Figure 4. Particle size distribution of sediment collected from
the reactor coolant drain tank.

Table V. Comparison of Reactor Building Air Cooler Radionuclide
Surface Concentrations with 305-ft-Elevation Vertical Metal
Radionuclide Surface Concentrations
(Data Decay-corrected to March 1, 1984)

Nuclide	Vertical Metal Surfaces[a] (μCi/cm^2)	Cooler External Surface (μCi/cm^2)	Cooler Internal Surface (μCi/cm^2)
Sr-90	1.6 ± 0.2 E-2	--	1.8 ± 1.7 E-2
I-129	4.2 ± 0.8 E-6	--	5.6 ± 2.2 E-6
Cs-134	1.6 ± 0.5 E-2	4.2 ± 2.4 E-2	4.3 ± 4.5 E-2
Cs-137	2.9 ± 1.0 E-1	7.8 ± 4.4 E-1	7.6 ± 8.0 E-1

a. McIsaac, C. V., "Surface Activity and Radiation Field Measurements of the TMI-2 Reactor Building Gross Decontamination Experiment," GEND-037, October 1983.

activities measured on other 305-ft-elevation vertical metal surfaces. For Sr-90 and I-129, average surface activities on the access
panels are in good agreement with average values for other vertical
metal surfaces at the 305-ft elevation. Average surface concentrations of Cs-134 and Cs-137 on the access panels appear to be about
2-1/2 times their corresponding concentrations on 305-ft-elevation
vertical metal surfaces, although the former values are statistically
in agreement with the latter values. It is important to note that
if we exclude the maximum values observed on the 11C and 11E access
panels, the average access panel internal surface concentration of
Cs-137 becomes 0.28 ± 0.16 $\mu Ci/cm^2$, which is in excellent
agreement with the corresponding 305-ft-elevation metal surface
values of 0.29 ± 0.10 $\mu Ci/cm^2$. This suggests that conditions
within the 11A, 11B, and 11D air coolers were unlike conditions in
the 11C and 11E air coolers. Surface activity concentrations on the
11C and 11E access panels are as much as five times greater than the
average 305-ft-elevation metal surface values.

Results of the in-situ gamma spectral measurements of the Reactor Building cooling coils and drip pans are presented in Table VI.
The scans were completed in October 1981 by Science Applications,
Inc., but the results are decay-corrected to March 1, 1984, in the
table to allow direct comparison of the results. Surface activity
values in $\mu Ci/cm^2$ were obtained by dividing the reported total
activities on the air cooler coils by the calculated surface area of
the air cooler coil. Calculated surface activity values for Cs-137
range from 0.059 ± 0.029 $\mu Ci/cm^2$ on air cooler coil 11E to
0.43 ± 0.07 $\mu Ci/cm^2$ on air cooler coil 11C. Calculated surface activity values for Cs-137 on the 11C and 11D cooling coils are
within a factor of 2.5 of the Cs-137 surface activities measured on
the corresponding access panel internal surfaces. The Cs-137 surface activity calculated for the 11E cooling coils is a factor of 36
lower than its average value on the corresponding access panel
internal of the 11E access panel. The lower activity concentrations
on the cooling coils, compared to the concentrations on the access
panels, are likely due to the cleansing effect of the water vapor
that condensed on the coils.

Surface activity concentrations of Cs-137 on the air cooler drip
pans are also reported in Table VI. These values range from 0.23 ±
0.12 $\mu Ci/cm^2$ on drip pan 11D to 0.81 ± 0.35 $\mu Ci/cm^2$ on drip
pan 11C. The surface activity values for Cs-137 on the 11C and 11D
air cooler drip pans are in excellent agreement with the concentrations of Cs-137 measured on the internal surfaces of the corresponding access panels, while the results for the 11E air cooler drip pan
are a factor of six lower than the average concentration of Cs-137
measured on the internal surface of the corresponding access panel.

Core Release Fractions. The total quantities of fission products,
fuel, and core materials measured to be present in the Reactor
Building, excluding the reactor core region and building atmosphere,
are summarized in this section.

The inventories of radionuclides and stable elements in the
basement water and sediment on each sampling date were calculated by
multiplying their measured concentrations times the volume of water
or the mass of sediment estimated to be present in the basement. The
volume of water in the basement on each sampling date is given in

Table II. The dry mass of the sediment in the basement is estimated
to be 380 kg. This estimate is based on an average sediment thickness of 0.635 cm and a sediment layer solids density of 63.5 mg/cm^3.
The inventories of a limited number of fission products in the RCS
were calculated by multiplying their measured concentrations times
the coolant volume of the RCS when at room temperature (21°C). The
RCS volume used was 3.33 x 10^5 L, where 4.16 x 10^4 L of the total
volume is attributed to the pressurizer. The calculated results are
expressed as percent of core inventory released. The core inventories of the fission products, fuel, and stable elements that were
considered in this summary are presented in Table VII.

Best estimates of the total quantities of fission products,
fuel, and core materials in the water and sediment in the Reactor
Building basement are summarized in Table VIII. The average quantities in solution are presented separately for 1979 and 1981. The
uncertainties in the quantities presented for the sediment and water
are, in each case, simply the standard deviation of the set of
measured concentration values. The uncertainties presented in
Table VIII do not include the uncertainties in the mass of sediment
and volume of water in the basement, nor do they include the uncertainties in the fission product inventories of the core that were
used to calculate the release fractions. The best estimates of the
total quantities in the basement are summations of the quantities in
the sediment and in solution. The results, expressed as percent of
core inventory, indicate that 57% of the H-3, 41% of the Cs-137, 20%
of the I-131, 14% of the I-129, and 1.7% of the Sr-90 were dispersed
in the water and sediment in the Reactor Building basement. The
uncertainty in the quantity of I-129 in the basement is at a minimum
±9% of the core inventory. This large uncertainty is due to the
fact that the concentration of I-129 in the samples of sediment
varied over a wide range of values.

The total quantities, expressed as percent of original core
inventory, of fission products, fuel, and core materials in the RCDT,
in the RCS water, and on the surfaces of the air cooling assembly and
reactor building structures are summarized along with the results for
the basement in Table IX. Measurements made on samples collected
from August 1979 to December 1983 have accounted for 59% of the H-3,
2.7% of the Sr-90, 15% of the I-129, 20% of the I-131, and 42% of the
Cs-137 originally in the core at the time of the accident.

The data summarized in Table IX show that the quantity of I-129
on Reactor Building contained surfaces is small. The quantity on
structural surfaces is about 0.06% of the original core inventory and
the amount estimated to be on the air cooling assembly surfaces is
about 0.2% of the original core inventory. Measurements made on
samples of Reactor Building air collected beginning 79 hours after
the onset of the accident have shown that the quantities of airborne
iodine were very small, ranging from 0.002 to 0.03% of the core
inventory (12).

Conclusions

Below are listed some of the conclusions of the radionuclide and
stable element measurements that have been performed on samples collected from the TMI-2 Reactor Building during the period from
August 1979 through December 1983.

Table VI. Reactor Building Air Cooler Radionuclide Surface
Concentrations[a]   (Data Decay-corrected to March 1, 1984)

Air Cooler Number	Cs-137 on Cooling Coils		Cs-137 on Drip Pans	
	(total Ci)	$(\mu Ci/cm^2)$[b]	(total Ci)	$(\mu Ci/cm^2)$[c]
11C	5.6 ± 1.0 E+0	4.3 ± 0.7 E-1	6.6 ± 2.8 E-2	8.1 ± 3.5 E-1
11D	1.5 ± 0.4 E+0	1.2 ± 0.3 E-1	1.9 ± 1.0 E-2	2.3 ± 1.2 E-1
11E	7.6 ± 4.0 E-1	5.9 ± 2.9 E-2	2.8 ± 8 E-2	3.5 ± 3.5 E-1

a. Quoted uncertainties include statistical uncertainties, estimates
of the uncertainties in the detector efficiency calibration, and
uncertainties in the data analysis.
b. Assumes surface area of coil ∿1.3 E+7 cm^2.
c. Assumes surface area of drip pans ∿8.2 E+4 cm^2.

Table VII. Core Inventory of Radionuclies and Elements
(Activities Decay-corrected to August 28, 1979)

Nuclide	Total Activity[a] (Ci)	Element	Total Mass[b] (kg)	Weight Percent[c]
H-3	3.65 E+3	Cr	5.71 E+2	0.47
Sr-90	7.68 E+5	Fe	1.42 E+3	1.15
Ru-106	2.42 E+6	Ni	7.67 E+2	0.62
Sb-125	5.00 E+4	Zr	2.28 E+4	18.59
I-129	1.95 E-1	Ag	2.20 E+3	1.79
I-131	1.29 E+1	In	4.12 E+2	0.34
Cs-134	1.86 E+5	Cd	1.37 E+2	0.11
Cs-137	8.37 E+5	Sn	3.34 E+2	0.27
Ce-144	1.62 E+7	Gd	3.40 E+0	0.003
		U	8.17 E+4[a]	66.81
		Pu	1.59 E+2[a]	0.13

a. Daniel, J. A.; Schlomer, E. A., Science Applications Inc.,
personal communication to G. R. Eidam, Bechtel National, Inc., TMI,
March 9, 1983.

b. Nuclear Associates International, "TMI-2 Accident Core Heatup
Analysis, Part III--As-built Design and Material Characteristics of
the TMI-2 Core," NSAC-25, June 1981.

c. Based on a total materials mass of 122, 737 kg.

Table VIII.   Best Estimates of Total Quantities of Fission Products, Fuel, and Core Materials in the Water Sediment in the Reactor Building Basement

		Percent of Core Inventory		
		Water[a]		
Isotope/ Element	Sediment[a] (1979 to 1983)	(1979)	(1981)	"Best Estimate" Total
H-3	--	5.7 ± 0.2 E+1	4.30 ± 0.07 E+1	5.7 ± 0.2 E+1
Sr-90	6.1 ± 7.2 E-2	6.6 ± 0.7 E-1	1.62 ± 0.07 E+0	1.7 ± 0.1 E+0
Ru-106	3.3 ± 2.6 E-3	4.3 ± 1.8 E-4	--	4 ± 3 E-3
Sb-125	1.2 ± 1.4 E-1	7.9 ± 2.8 E-2	2.1 ± 0.6 E-1	3 ± 2 E-1
I-129	7.6 ± 8.7 E+0	1.38 ± 0.04 E+1	5.9 ± 0.9 E+0	1.4 ± 0.9 E+1
I-131	1.6 ± 0.4 E+0	1.86 ± 0.09 E+1	--	2.0 ± 0.1 E+1
Cs-134	4.6 ± 5.7 E-2	3.96 ± 0.32 E+1	4.16 ± 0.16 E+1	4.2 ± 0.2 E+1
Cs-137	4.2 ± 4.9 E-2	4.01 ± 0.13 E+1	4.08 ± 0.12 E+1	4.1 ± 0.1 E+1
Ce-144	9.1 ± 5.7 E-4	4.5 ± 4.5 E-5	<3.3 E-3	1.0 ± 0.6 E-3
U	8.0 ± 6.1 E-4	3.8 ± 2.6 E-5	<8.7 E-5	8 ± 6 E-4
Pu	6.1 ± 5.3 E-4	1.8 ± 1.3 E-5	3.2 ± 1.0 E-4	6 ± 5 E-4
Cr	3.5 ± 3.8 E-1	2.4 ± 0.0 E-1	1.1 ± 0.3 E-1	5 ± 4 E-1
Fe	1.1 ± 1.0 E+0	1.6 ± 0.8 E-1	1.4 ± 0.2 E-1	1 ± 1 E+0
Ni	7.6 ± 6.1 E-1	<2.5 E-1	3.4 E-1	8 ± 6 E-1
Zr	3.6 ± 1.5 E-4	--	1.5 E-2	4 ± 2 E-4
Ag	1.9 ± 2.1 E-1	<2.7 E-2	<1.1 E-1	2 ± 2 E-1
Cd	2.9 ± 2.7 E-1	<2.9 E-2	<3.5 E+0	3 ± 3 E-1
In	8.7 ± 13 E-1	<4.7 E-2	<2.9 E+0	9 ± 13 E-1
Sn	7.2 ± 6.5 E-2	--	<3.6 E+0	7 ± 6 E-2
Gd	<1.1 E-1	--	<7.0 E+1	<1 E-1

a.   Uncertainty is given at the one-sigma level and is the standard deviation of the set of concentration values.

Table IX. Total Quantities of Fission Products, Fuel, and Core
Materials in the TMI-2 Reactor Building Expressed As
Percent of Original Core Inventory

Isotope/ Element	Basement	Reactor Coolant Drain Tank	Reactor Coolant System	Air Cooling Assembly	Structural Surfaces[a]	Total
H-3	5.7 E+1	3.3 E-2	2.2 E+0	--b	--b	5.9 E+1
Sr-90	1.7 E+0	6.3 E-2	9.6 E-1	2.0 E-4	2.3 E-3	2.7 E+0
Ru-106	4 E-3	1.6 E-3	--b	--b	--b	5.6 E-3
Sb-125	3 E-1	3.6 E-3	--b	--b	1.1 E-2	3.1 E-1
I-129	1.4 E+1	5.5 E-3	1.2 E+0c	2.2 E-1	6.4 E-2	1.5 E+1
I-131	2.0 E+1	--b	--b	--b	--b	2.0 E+1
Cs-134	4.2 E+1	5.2 E-3	7.7 E-1	8.1 E-3	3.5 E-2	4.3 E+1
Cs-137	4.1 E+1	4.7 E-3	8.1 E-1	7.7 E-3	3.5 E-2	4.2 E+1
Ce-144	1.0 E-3	9.5 E-4	--b	--b	--b	2.0 E-3
U	8 E-4	1.2 E-4	--b	--b	--b	9.2 E-4
Pu	6 E-4	--b	--b	--b	--b	6 E-4
Cr	5 E-1	1.2 E-2	--b	--b	--b	5.1 E-1
Fe	1 E+0	2.1 E-1	--b	--b	--b	1.2 E+0
Ni	8 E-1	3.8 E-1	--b	--b	--b	1.2 E+0
Zr	4 E-4	2.2 E-4	--b	--b	--b	6.2 E-4
Ag	2 E-1	7.4 E-3	--b	--b	--b	2.1 E-1
Cd	3 E-1	2.3 E-1	--b	--b	--b	5.3 E-1
In	9 E-1	4.0 E-2	--b	--b	--b	9.4 E-1
Sn	7 E-2	3.9 E-2	--b	--b	--b	1.1 E-1
Gd	<1 E-1	1.5 E-1	--b	--b	--b	1.5 E-1

a. Total surface activities were reported in "Surface Activity and
Radiation Field Measurements of the TMI-2 Reactor Building Gross
Decontamination Experiments," GEND-037, October 1983, p. 102.

b. Not measured.

c. August 14, 1980 reactor coolant system coolant sample.

- With the exception of Sr-90 and Ce-144, the vast majority of
  the total quantity of each radionuclide released from the core
  was found dispersed in the water and sediment in the basement.
- The quantities of fission products on upper-level building
  surfaces were each less than 0.3% of their original core
  inventories.
- About 57% of the I-129 that was in solution in the basement
  water in August 1979 was lost from the water by March 1981.
  Based on the measured average concentration, the total amount
  of I-129 in the sediment on the basement floor is about equal
  to the amount of I-129 that was lost from solution. This
  indicates that precipitation was probably the dominant
  depletion mechanism.
- The data indicate a good correlation between the concentrations
  of Cu and Ag and the concentration of I-129 in the sediment on
  the basement floor. This suggests that iodine precipitated as
  Cu and Ag iodide and/or iodate.

- The relative quantities of control rod alloy and core structural material elements that were measured in the particulate matter in the RCDT correlate well with their calculated relative volatilities. This indicates that non-fuel metals were transported to the RCDT as condensed vapors.

- The quantity of radioiodine, expressed as percent of original core inventory, that has been accounted for outside the core region inside the Reactor Building is less than one-half the amount of radiocesium that has been accounted for. Possible sinks for the missing iodine include (a) the letdown demineralizers, (b) the letdown heat exchangers, (c) sediment on the basement floor, and (d) RCS surfaces.

Acknowledgments. This work was supported by the U.S. Department of Energy, Assistant Secretary for Nuclear Energy, Office of Remedial Action and Terminal Waste Disposal, under DOE Contract No. DE-ACO7-76ID01570.

Literature Cited

1. Croucher, D. W. "Three Mile Island Unit 2 Core Status Summary: A Basis for Tool Development for Reactor Disassembly and Defueling," GEND-007, May 1981.
2. TMI-2 Technical Planning Department, GPU Nuclear-Bechtel National, Inc., "Reactor Building Basement—History and Present Conditions," TPO/TMI-027, Rev. 0, November 1982.
3. McIsaac, C. V. "Surface Activity and Radiation Field Measurements of the TMI-2 Reactor Building Gross Decontamination Experiment," GEND-037, October 1983.
4. Runion, T. C., personal communication to Holzworth, R. E., EG&G Idaho, Inc., April 2, 1981.
5. Campbell, D. O.; Malinauskas, A. P., Oak Ridge National Laboratory, personal communication to G. R. Eidam, EG&G Idaho, Inc., November 13, 1981.
6. TMI-2 Technical Planning Department, GPU Nuclear-Bechtel National, Inc., "Reactor Coolant System Sample Results," TPO/TMI-122, Rev. 0, July 1984.
7. Willis, C. P., personal communication, EG&G Idaho, Inc., September 25, 1984.
8. Nuclear Associates International, "TMI-2 Accident Core Heatup Analysis, Part III—As Built Design and Material Characteristics of the TMI-2 Core," NSAC-25, June 1981.
9. Petty, D., private communication, Massachusetts Institute of Technology Graduate School, Cambridge, MA, August 28, 1984.
10. Akers, D. W. Idaho National Engineering Laboratory, personal communication to McIsaac, C. V., EG&G Idaho, Inc., March 15, 1985.
11. Cubicciotti, D. and Sehgal, B. R., "Vaporization of Core Materials in Postulated Severe Light Water Reactor Accidents," Nuclear Technology, 1984, 67.
12. Pelletier, C. A. et al., "Preliminary Radioactive Source Term and Inventory Assessment for TMI-2," GEND-028, March 1983.
13. R. C. Weast (ed.), "Handbook of Chemistry and Physics," 52nd edition, CRC Press, 1971-1972, pp. B164-165.

RECEIVED September 25, 1985

Iodine Chemistry

J. Paquette, D. J. Wren, and B. L. Ford

Research Chemistry Branch, Atomic Energy of Canada Limited, Whiteshell Nuclear
Research Establishment, Pinawa, Manitoba, Canada ROE 1LO

This paper describes the main features of the chemistry
of iodine that contributed to the low concentration of
radioactive iodine observed in the gas phase following
the Three Mile Island-2 nuclear reactor accident. The
very low concentration of iodine in the gas phase was
one of the most important and, to some, surprising
feature of the accident. The behaviour of radioactive
fission products, such as iodine, during a loss-of-
coolant accident is a complex function of their
chemistry in UO_2 fuel, in the gas phase and in the
aqueous phase. The main focus of this paper is on the
chemistry of iodine in aqueous solution, including
thermodynamic and kinetic calculations of aqueous iodine
reactions, such as hydrolysis, disproportionation,
reaction with organic material and radiolysis. From an
examination of the chemistry of iodine in nuclear fuel,
in the reactor coolant system, and in the containment
building, it is concluded that the chemical conditions
at Three Mile Island-2 favoured low iodine volatility.

Following a severe nuclear reactor accident in which fuel failure
occurs, radioactive material could be released to the environment by
leakage through the containment building and from filtered-air discharge systems. Of all the fission products that can be released
from a nuclear reactor core to the environment, the radioactive
isotopes of iodine, tellurium, bromine, krypton and xenon are considered to be the most important ones. Isotopes of tellurium and
bromine are important insofar as they are precursors to iodine and
krypton. Iodine and the noble gases are considered to be important,
not only because of a combination of inventories and half-lives but
also, because their potential volatility makes them difficult to
contain.
    The noble gases are chemically and biologically inert, whereas
iodine is chemically reactive and biologically active. For these

0097–6156/86/0293–0193$06.00/0
© 1986 American Chemical Society

reasons, iodine has been considered, in the past, as the limiting
radionuclide in terms of radiation exposure to the public in the
analysis of nuclear reactor accident consequences. Such analyses
have usually assumed that a large fraction of the iodine released
from defected nuclear fuel would become airborne in the containment
building and be readily available for release to the environment
(1).

In March 1979, unit 2 at Three Mile Island, a 900 MW(e) pressurized water reactor, suffered a partially mitigated loss-of-
coolant accident, which eventually led to severe core damage.
Extensive fission-product release occurred, and about 40% of the
radioiodine present in the reactor core was dispersed in the
containment and auxiliary buildings. One of the most important and,
to some, surprising features of the accident was the very low
concentration of iodine found subsequently in the gas phase of the
containment building. An understanding of how this occurred could
lead to more realistic analyses of the radiological consequences of
nuclear reactor accidents. More importantly, it could help identify
processes that could be effective for the abatement of iodine
following such accidents.

The fate of iodine following its accidental release from a
reactor core depends on a variety of physical and chemical factors.
In this paper, we examine the main features of the chemistry of
iodine that could have contributed to the low volatility of this
element at Three Mile Island-2. We discuss, in turn, the chemistry
of iodine in nuclear fuel, in the reactor coolant system, and in the
containment building. The main focus is on the thermodynamics and
kinetics of these systems.

Iodine Chemistry in Fuel

An examination of the behaviour of iodine following a reactor accident should start with the chemistry of iodine in UO_2 fuel. Iodine
isotopes in fuel are formed mostly by the beta decay of tellurium
isotopes, except for ^{135}I, which is also formed directly by fission.
Iodine is believed to migrate as an atom to microbubbles or fis-
sures, where it may react to form I_2 or other species such as CsI.

The current understanding is that iodine that has escaped the
UO_2 lattice combines with the fission product cesium to form CsI
(2,3). This is based on equilibrium thermodynamic calculations by
Besmann and Lindemer (4), which show that CsI is the most stable
iodine species in the Cs-U-Zr-H-I-O chemical system at oxygen
partial pressures close to that of $UO_{2.0}$. Cubicciotti and Sanecki
(5) have observed CsI deposits on the inner surfaces of fuel
cladding. Studies by Lorentz et al. (6) of fission-product release
from irradiated fuel fragments at temperatures above 700°C have
provided indirect evidence that iodine is released as CsI. Studies
of iodine and cesium within an irradiated fuel pellet showed that
both have similar radial concentration profiles, but showed no clear
formation of CsI phases (7). The above evidence is consistent with
CsI formation within the fuel, but is not sufficient to determine
the chemical state of iodine within the fuel unambiguously.

During the accident at Three Mile Island-2, iodine was released
from fuel elements to the reactor coolant system in two stages.

When the upper part of the fuel elements was first uncovered and the
temperature rose above 700°C, the elements began to fail. This
resulted in the release of the gap inventory of iodine, almost
certainly as CsI. The fraction of the core inventory of iodine
released during this phase was of the order of 0.04% (8). The rest
of the iodine was released during the heating of the fuel elements,
as the water level in the core slowly dropped and a Zircaloy-steam
reaction occurred. During this time, iodine and cesium were
probably released to the reactor coolant system as atoms, due to
high temperatures in the core region.

Iodine Chemistry in the Reactor Coolant System

Thermodynamics. Some areas of the reactor coolant system contained
only steam and hydrogen during the accident at Three Mile Island-2,
while other areas contained steam and liquid water. The region
above the uncovered fuel elements and the lower plenum assembly was
subjected to the highest temperatures and contained the largest
volume of steam and hydrogen (9).

The equilibrium speciation of iodine in such a high-temperature
gaseous system can be calculated using the principles of chemical
thermodynamics. Several groups have calculated the equilibrium
speciation in Cs-I-H-O systems (10-13). These calculations indicate
that CsI is the dominant stable species at lower temperatures, while
at higher temperatures, CsI is less stable and I and HI become
important. The relative stability of CsI and the temperature at
which the changeover occurs depend on a number of factors, including
total pressure, iodine concentration, Cs:I ratio, and H:O ratio.
Figures 1 and 2 give the equilibrium iodine speciation for different
conditions. These figures show that the factors that favour CsI at
higher temperatures are high iodine concentration, high Cs:I ratio
and high H:O ratio. The reactor at Three Mile Island-2 was not
instrumented to measure these parameters, but we can infer some of
these from post-accident analyses (9). The extensive oxidation of
the Zircaloy cladding caused considerable H_2 production, and the
atmosphere in the plenum was certainly reducing with H:O > 2. Also,
a large fraction of both the iodine and the cesium were released
from the fuel. Since the Cs:I ratio is ∿ 10 in the fuel, and both I
and Cs are released at the same rate from overheated fuel, CsI
formation was further favoured. Finally, post-accident thermal-
hydraulic analyses provide an estimate of the gas temperatures in
the upper plenum region. Ardron and Cain (9) have estimated that
the gas temperatures in the upper plenum never exceeded 1575 K and
rarely exceeded 1175 K during the accident. From an examination of
Figures 1 and 2, it is clear that CsI is the thermodynamically
stable, airborne iodine species under these conditions. Furthermore, as iodine was swept from the upper plenum into cooler regions
of the reactor coolant system, CsI formation would have been
increasingly favoured.

Kinetics. The formation of CsI in the reactor coolant system, while
thermodynamically favoured, is not assured unless it is also
kinetically allowed. This is particularly true in a steam environment, because most of the cesium atoms released could initially
react to form CsOH:

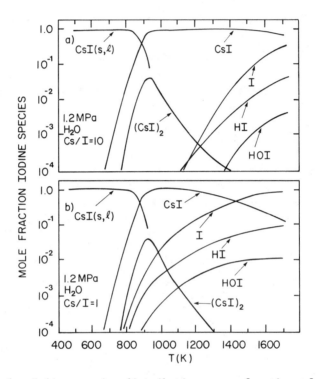

Figure 1.   Iodine species distributions as a function of
temperature for a steam atmosphere at 1.2 MPa containing 25 μg/g
total iodine: (a) Cs/I = 10, and (b) Cs/I = 1.

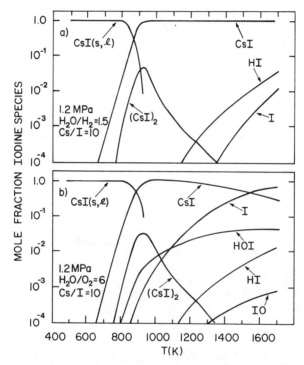

Figure 2. Iodine species distributions as a function of temperature at 1.2 MPa for two systems containing 25 μg/g total iodine and Cs/I = 10: (a) steam and hydrogen atmosphere (steam/hydrogen = 1.5), and (b) steam and oxygen (air) atmosphere (steam/oxygen = 6).

$$Cs + H_2O \rightarrow CsOH + H \tag{1}$$

Subsequent reactions with iodine atoms, such as

$$CsOH + I \rightarrow CsI + OH \tag{2}$$

would have to occur fast enough to ensure CsI formation. Figures 3 and 4 show the results of two kinetic calculations on the change in speciation with time following release of Cs and I atoms, and of $CsOH$ and I atoms, into a steam/hydrogen atmosphere at 1000 K. In both cases, CsI becomes the dominant form of iodine in less than 0.01 s. Thus, CsI formation above the fuel elements and in the lower plenum was both kinetically and thermodynamically favoured at Three Mile Island-2.

Specific Reactions. The above calculations also show that, in a hydrogen-rich atmosphere, any iodine not bound to cesium will be present mostly as HI. This would be particularly true for iodine released early, or late, in an accident process, when the total concentration of airborne Cs and I in the upper plenum is low and CsI is not thermodynamically favoured because of a shift in the equilibria

$$CsI + H_2O \rightleftharpoons CsOH + HI \tag{3}$$
and
$$I_2 + H_2 \rightleftharpoons 2HI \tag{4}$$

Even though HI is a reactive gas, in the Three Mile Island-2 case, the steam and hydrogen flow rates from the core during the boil-off time should have carried any HI through the reactor coolant system rapidly and prevented reactions with steel surfaces. On contact with liquid water in the cooler part of the coolant system, HI would have dissociated completely to give I^-(aq):

$$HI(aq) \rightarrow H^+(aq) + I^-(aq) \tag{5}$$

The airborne CsI formed in the upper plenum of the reactor should also have been transported away from the reactor core area. However, due to the high melting point of CsI (621°C), it would have condensed on relatively cold surfaces, or aerosol particles (some of which could be formed by self-condensation of CsI). Cesium iodide can react chemically with surfaces, however, experiments by Elrich and Sallach (14) have shown that the reaction of CsI with stainless steel, to form I_2, is slow. Hence, this was probably unimportant at TMI-2. There is also evidence that CsI may react with solid borates (14), e.g.:

$$CsI + HBO_2 \rightarrow CsBO_2 + HI \tag{6}$$

Boron was present in the TMI-2 coolant as an additive to control the core reactivity. However, since there was always some water in the core and the reactor coolant system during the accident, this reaction was probably not important.

Hahn and Ache (12) have pointed out that silver released from the control rods at Three Mile Island-2 may have played an important

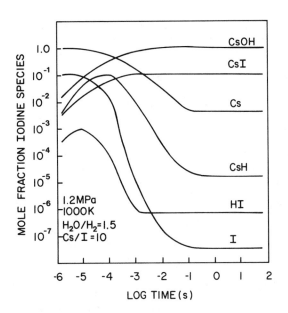

Figure 3. Iodine species distribution as a function of time after release of Cs and I atoms (Cs/I = 10 and 30 μg/g total iodine) into a 1.2 MPa steam/hydrogen atmosphere at 1000 K.

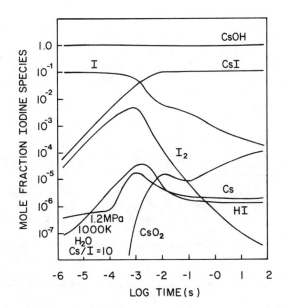

Figure 4. Iodine species distribution as a function of time after release of CsOH and I (CsOH/I = 10 and 30 μg/g total iodine) into a 1.2 MPa steam atmosphere at 1000 K.

role. There was approximately 120 times as much silver in the
control rods as there was iodine in the reactor core. These authors
have shown that, even if only a fraction of this silver was
released, the dominant form of iodine could have been AgI rather
than CsI. This depends on whether the reaction

$$Ag + CsI \rightarrow AgI + Cs \tag{7}$$

is kinetically favoured or not.

Mitigating against AgI formation are the different release
rates for Cs and I compared with Ag. In an accident situation, the
Cs and I would be released continuously following an initial gap
release, and a substantial fraction would be released before the
stainless-steel clad control rods failed and significant amounts of
silver were released. Temperatures greater than $1600°C$ would be
required for the latter to occur. For a slow core heat-up, such as
that which occurred at TMI-2, much of the CsI probably left the fuel
and the plenum before the Ag release began. There may still have
been some AgI formation. Post-accident analyses found 51 to 68% of
the inventory of Cs released from the core, but only 36% of the
inventory of iodine (15). The difference may be due to iodine that
reacted to form insoluble AgI, and which was retained in the core
and in the reactor coolant system.
    Unreacted CsI was transported to water in the cooler part of
the reactor coolant system, where it would have dissolved and dissociated to form I . Oxidation of dissolved I , either chemically
or through radiolysis, was unlikely since dissolved hydrogen was
most certainly present in the reactor coolant. Hydrogen is a rapid
and effective reducing agent in water, when the temperature is above
$150°C$, especially in the presence of a radiation field. Moreover,
hydrazine was present in the reactor coolant for corrosion preven-
tion. Hydrazine is a powerful reducing agent and reacts rapidly
with iodine in aqueous media; it also decomposes to hydrogen and
ammonia in a radiation field. The process of transporting CsI
through the reactor coolant system may have involved condensation
and revaporization, but the end result was that iodine released into
the reactor containment or auxiliary building was predominantly in
the form of I dissolved in water.

Iodine Chemistry in the Containment Building

Thermodynamics. Large quantities of reactor coolant were released
to the containment and auxiliary buildings at Three Mile Island-2.
Under those conditions, iodine would have eventually partitioned
between the aqueous phase and the gaseous phase. The equilibrium
volatility of iodine, for a gas phase in contact with an aqueous
phase, in a closed system, can be calculated using the principles of
chemical thermodynamics, and is independent of the initial chemical
state of iodine. This calculation requires a knowledge of the
equilibrium speciation of iodine in solution. Past studies have
revealed that the iodine/water system is complex, since the element
can exist in various oxidation states and is subject to hydrolysis
and disproportionation in aqueous media. Despite this complexity,
the thermodynamics of the system are well understood, owing to the

efforts of a large number of researchers over the last sixty years.
The recent literature data for aqueous iodine species are, but for a
few exceptions, sufficiently accurate.

The equilibrium behaviour of aqueous systems usually depends
strongly on pH, electrochemical potential and temperature. A
convenient way of representing the behaviour of the iodine/water
system for various pH and potential conditions is by using
potential/pH diagrams. The thermodynamic values tabulated in
reference (16) and the computational techniques described in
reference (17) where used to construct the diagram shown in Figure
5. Such a diagram provides a guide to the chemistry of the system
by showing the predominant species in aqueous media for various
potential and pH conditions. Figure 5 indicates, that if the pH is
above a value of about 3, the predominant species are iodide under
reducing conditions and iodate under oxidizing conditions. It is
interesting to note that both of these species, being ionic, are
non-volatile.

A more detailed equilibrium calculation can be performed for a
given redox condition at a fixed pH. The results of such a calculation are illustrated in Figure 6, which shows the concentration of
various iodine species as a function of the electrochemical
potential at a pH value of 8, for a total iodine concentration of
10^{-8} mol·dm^{-3}. Figure 6 demonstrates that strongly reducing and
strongly oxidizing conditions would minimize the equilibrium volatility of iodine by stabilizing the non-volatile iodide or iodate
ions, as could be expected from the more general potential/pH calculation. Mildly reducing conditions, at near neutral pH, would tend
to maximize the concentration of the potentially volatile HOI and I_2
species. It is worth noting that, at low total iodine concentration, the equilibrium concentration of elemental iodine is always at
least four orders of magnitude less than that of HOI, which is, in
turn, always lower by at least two orders of magnitude than the
concentration of either I^- or IO_3^-.

A calculation of the equilibrium partition coefficient (concentration of iodine in the aqueous phase/concentration of iodine in
the gaseous phase) requires values for the volatility of the aqueous
iodine species. The ionic species can be considered as non-
volatile. Thus, the potentially volatile iodine aqueous species are
HIO_3, HOI and I_2. There is no report in the literature on the
volatility of HIO_3, and we have not been able to detect HIO_3 in the
gas phase above an acid aqueous solution of HIO_3. The volatility of
HIO_3 is expected to be much less than that of I_2 and can thus be
ignored. Recent experiments in our laboratory (18) and at Oak Ridge
National Laboratory (19) have indicated that HOI is far less volatile than was believed in the past. According to these recent
studies, an upper limit for the equilibrium constant for the
reaction

$$HOI(aq) \rightleftharpoons HOI(g) \tag{8}$$

is $< 1\mathrm{x}10^{-3}$ atm·dm^3·mol^{-1}, which amounts to a HOI partition coefficient $> 10^4$. The volatility of I_2 has been determined several times
in the past, and its value is known accurately for the temperature
range 0 to 100°C.

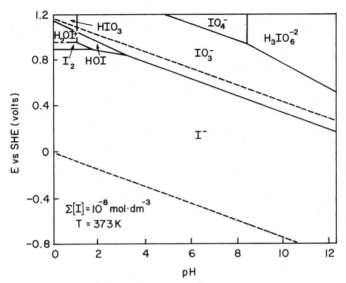

Figure 5. Potential-pH diagram for iodine in water at 100°C, for a total iodine concentration of 10^{-8} mol·dm^{-3}.

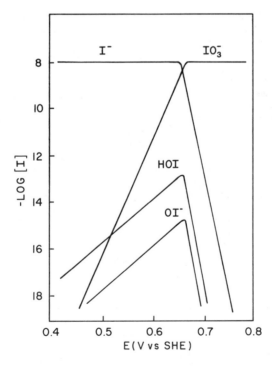

Figure 6. Distribution diagram for iodine in water at 60°C, for a pH value of 8 and a total iodine concentration of 10^{-8} mol·dm^{-3}.

Using the above equilibrium aqueous speciation and the known
volatilities, the overall partition coefficient was calculated as a
function of the electrochemical potential at a pH value of 8. This
is illustrated in Figure 7. This Figure shows that the equilibrium
partition coefficient goes through a minimum as the conditions
change from very reducing to very oxidizing. This is because
reducing and oxidizing conditions favour the non-volatile species I^-
and IO_3^-, respectively. On the other hand, mildly reducing conditions favour the more volatile HOI and I_2 species. Figure 8 is a
plot of the partition coefficient as a function of pH for a fixed
electrochemical potential. The partition coefficient increases with
the pH, due to I^- and IO_3^- becoming more and more favoured over HOI
and I_2 as the pH rises. It is obvious that both the potential and
the pH can have a major impact on the partition coefficient of the
iodine/water system. A high pH would tend to maximize the partition
coefficient (i.e., iodine would favour the aqueous phase). Strongly
reducing or strongly oxidizing conditions would also maximize the
partition coefficient, whereas mildly reducing conditions would
lower the partition coefficient (i.e., iodine would favour the gas
phase).

In a nuclear reactor accident, such as at Three Mile Island-2,
the potential and the pH would not be controlled by the small amount
of iodine present. The containment building water would contain, in
addition to debris, substantial amounts of dissolved fission
products, as well as various organic contaminants and dissolved
ions, which could affect the pH and the potential in ways difficult
to predict accurately. At Three Mile Island-2, during and immediately after the accident, boric acid and sodium hydroxide were added
to the primary coolant. Hydrazine was also added to maintain the
dissolved oxygen level below 100 mg/g. Hydrazine is a strong
reducing agent that can scavenge oxygen and directly reduce iodine.
It is decomposed by strong radiation fields to produce hydrogen,
which is an effective oxygen scavenger in a high radiation field.
The containment spray system, which contains sodium hydroxide, was
also initiated by a hydrogen burn later in the accident sequence.
Thus, the conditions were certainly basic and reducing which, from
an equilibrium viewpoint, would minimize the volatility of iodine.

The inclusion of possible reactions between iodine species and
organic compounds does not affect the above thermodynamic analysis
significantly. Although reactions such as:

$$HOI(aq) + RH(aq) \rightleftharpoons RI(aq) + H_2O \qquad (9)$$

are thermodynamically favoured, the resulting organic iodides are
themselves unstable towards hydrolysis:

$$RI(aq) + H_2O \rightleftharpoons I^- + ROH(aq) + H^+ \qquad (10)$$

A thermodynamic analysis provides a guide to the possible
extent of iodine reactions under a variety of chemical conditions.
However, it provides no information on the behaviour of the system
prior to equilibrium. If chemical reaction rates are slow, then the
chemical forms of iodine for the time period of interest may be
markedly different from those anticipated at equilibrium. For
example, if at equilibrium the vapour phase contains even 1% by

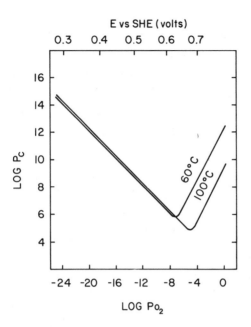

Figure 7. Partition coefficient as a function of the electrochemical potential for iodine in water at a pH value of 8.

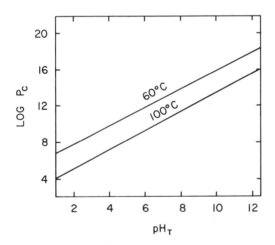

Figure 8. Partition coefficient as a function of pH for iodine in water at an electrochemical potential of 0.750 V vs SHE.

volume of oxygen, then 99.9% of iodine, in any initial form, will be converted to iodate, and the amount of iodine in the gas phase will be very small. However, this behaviour is relevant to an accident situation only if iodine is in a form that can be rapidly converted to iodate. Thus, for a complete description of the chemistry of the system, kinetic factors have to be considered.

Kinetics. Kinetic analyses are inherently more difficult and less accurate than equilibrium calculations. There is no theory that allows the prediction of the kinetic behaviour of an aqueous system from a few key measurements. Reaction rates are highly specific to the species involved and can vary widely for a given reaction, depending on the chemical conditions.

It is likely that the iodine species released to the containment building during the Three Mile Island-2 accident was the iodide ion. Aqueous iodide itself is non-volatile and would have to be oxidized in order to form any volatile species. The major oxidizing agent in the containment building water at Three Mile Island-2 was most certainly dissolved atmospheric oxygen. In this context, the most important reaction sequence is the direct oxidation of iodide by dissolved oxygen:

$$2I^- + 2H^+ + 1/2\ O_2 \rightarrow I_2 + H_2O \qquad (11)$$

followed by the instantly established hydrolysis equilibrium:

$$I_2 + H_2O \rightleftharpoons HOI + I^- + H^+ \qquad (12)$$

Iodine is then transformed from an non-volatile, non-reactive form to a highly reactive and potentially volatile mixture of I_2 and HOI.

The uncatalyzed rate of iodide oxidation is extremely slow for low iodide concentrations near neutral pH, the rate law being (20)

$$-\frac{d[I^-]}{dt} = 1.3 \times 10^{-4} [O_2] [I^-] [H^+] \qquad (13)$$

in units of $mol \cdot dm^{-3} \cdot s^{-1}$ at 25°C. Recent studies by Burns and Marsh (21) have demonstrated that the oxidation is slow even at 300°C. The reaction is known to be catalyzed by light and by the presence of metal ions such as Cu^{+2} and Fe^{+3} in trace amounts. However, even the catalyzed reaction is relatively slow.

Any iodide that is oxidized would be subject to disproportionation according to

$$3I_2 + 3H_2O \rightleftharpoons 5I^- + IO_3^- + 6H^+ \qquad (14)$$

This reaction, as opposed to the oxidation of I^-, involves the conversion of a reactive and volatile species into a non-reactive, non-volatile mixture. The overall reaction (14) is, in fact, a combination of (12) and of the reaction of the intermediate +1 oxidation state species HOI:

$$3HOI \rightarrow 2I^- + IO_3^- + 3H^+ \qquad (15)$$

Since the formation of HOI from I_2 (reaction (12)) is rapid,
substantial amounts of HOI could accumulate in solution if reaction
(15) is slow. We have demonstrated in our laboratory (22) that the
reaction is second order in HOI and obeys the rate law

$$- \frac{d[HOI+I_2+I_3^-]}{dt} = k[HOI]^2 \qquad (16)$$

where k has a value of about 100 $dm^3 \cdot mol^{-1} \cdot s^{-1}$ near neutral pH at
25°C. This means that, for low concentrations of iodine, HOI can
accumulate and persist for long periods of time in solution. As an
example, its half-life for an initial concentration of
1×10^{-8} $mol \cdot dm^{-3}$ would be 12 days at a pH value of 8. Although HOI
is far less volatile than I_2, it could be more reactive towards
organic impurities, leading to organic iodide formation, as
discussed below.

Organic Iodides. One of the more perplexing aspects of the
behaviour of iodine following a serious reactor accident is the
formation of volatile organic iodides. Measurements at Three Mile
Island-2 have shown that, prior to venting, up to 90% of the
airborne iodine was present in an organic form (15). After venting,
the amount of organic iodide in the gas phase slowly returned to its
initial level.

Since most of the iodine at Three Mile Island-2 was in solution, it is worth examining the possible mechanism for organic
iodide formation in water containing various iodine species and
organic impurities. Hypoiodous acid, in particular, would be
expected to be highly reactive towards organic material, by analogy
with the homologous HOCl and HOBr. Elemental iodine itself is
reactive in solution, particularly with unsaturated hydrocarbons.
Iodination of phenol, secondary methyl ketones and alcohols, and
various large organic molecules occurs readily in solution.

We have investigated the rate of reaction of iodine solutions
with phenol and 2-propanol (23). Iodophenol was formed rapidly in
the first case, by reaction with both HOI and I_2. Triodomethane
(iodoform) was formed more slowly in the reaction with 2-propanol
and could be easily identified in the gas phase by mass spectrometry. Hypoiodous acid was found to be the reactive species.
Volatile organic species were also identified by mass spectrometry
following reaction between HOI and oil-contaminated, sump-pump water
samples. This clearly demonstrates that organic iodides can be
formed rapidly by reaction with organic contaminants and can be
transferred to the gas phase. Thus, the behaviour of volatile
iodides at Three Mile Island-2 is not inconsistent with a mechanism
involving solution reactions, although other mechanisms, such as
radiolysis, could also explain the observed phenomena.

Radiolysis. High radiation fields were present in the containment
building at Three Mile Island-2 immediately after the accident.
Radiolysis can have an important impact on the interconversion of
aqueous iodine species and must be considered for a complete
description of the system. The radiolysis of iodine solutions has

been studied extensively over the last twenty years, and the radiation-induced interconversion between I^- and I_2 is relatively well understood (21). The primary interaction of radiation is with water, the major component of the system. Radiolytic decomposition of water produces a mixture of highly reactive oxidizing and reducing species according to the general equation

$$H_2O \rightarrow OH + H + e^- + H_2 + H_2O_2 \qquad (17)$$

The radicals thus formed can attack the various iodine species present to form iodine radicals:

$$I^- + H + H^+ \rightarrow I + H_2 \qquad (18)$$

$$I^- + OH \rightarrow HOI^- \qquad (19)$$

$$I_2 + e^- \rightarrow I_2^- \qquad (20)$$

$$I_2 + H \rightarrow I_2^- + H^+ \qquad (21)$$

The iodine radicals can react further with water radiolysis products, or with the stable iodine species, or can recombine:

$$I_2^- + e^- \rightarrow 2I^- \qquad (22)$$

$$I_2^- + H \rightarrow 2I^- + H^+ \qquad (23)$$

$$I + I^- \rightarrow I_2^- \qquad (24)$$

$$IOH^- + I \rightarrow I_2^- + OH^- \qquad (25)$$

$$I + I \rightarrow I_2 \qquad (26)$$

$$I + I_2^- \rightarrow I_2 + I^- \qquad (27)$$

$$I_2^- + I_2^- \rightarrow I_2 + 2I^- \qquad (28)$$

Reactions with the stable products of water radiolysis such as H_2O_2 also occur, but at a much slower rate. Additional reactions occur in the presence of oxygen due to the formation of the radical species O_2^- and HO_2:

$$HO_2 + 2I^- + H^+ \rightarrow H_2O_2 + I_2^- \qquad (29)$$

$$O_2^- + I_2^- \rightarrow 2I^- + O_2 \qquad (30)$$

The net result of this complex group of reactions is that, for a solution containing I^-, oxidation will occur during the initial stage of radiolysis. However, since the radiolysis of water produces a mixture of oxidizing and reducing species, once appreciable concentrations of oxidized iodine species start to build up, iodide will be reformed by a series of reducing reactions. The balance of oxidation and reduction favours I^- at neutral pH, and increasingly so as the pH increases. The radiation-induced steady state leans strongly towards I^- at pH values from 7 to 10. This is even more true if dissolved hydrogen is present, as was the case at Three Mile Island-2, since dissolved hydrogen scavenges the oxidizing OH radicals to produce the reducing radicals H:

$$H_2 + OH \rightarrow H + H_2O \qquad (31)$$

The above conclusions are valid for pure iodide solutions. The picture can be modified considerably in the presence of impurities that can scavenge specifically either the reducing or the oxidizing species, thus affecting the redox balance. As an example, trace amounts of Cu^{+2} can shift the balance more toward I_2 (24), whereas borate appears to have the opposite effect. Similarly, Burns and Marsh (21) have observed that the presence of Teflon can increase iodine volatility substantially, while some metallic surfaces can produce the opposite effect. The containment building water at Three Mile Island-2 contained substantial amounts of dissolved fission products, organic contaminants, various dissolved ions, some particulate matter and debris. The role of these extraneous materials in influencing the radiation chemistry of iodine is not well known.

The radiation chemistry of aqueous iodine species in higher oxidation states is also poorly known. A large number of intermediate species and reactions, mostly hypothetical, have been proposed. Experimental results are often conflicting. There seems to be some agreement in the literature on the following points: (i) OH, the chief oxidizing agent resulting from water radiolysis, does not react with I_2, and (ii) e^- appears to reduce IO_3^-. There is no information on the radiolysis of the +1 oxidation state species HOI(aq), which is an important species in solution.

Organic material will also be affected by radiation. This could potentially lead to formation of organic iodides according to the general reaction scheme:

$$RH + radical \rightarrow \ ^\cdot R + H-radical \qquad (32)$$

$$^\cdot R + I-compound \rightarrow RI + \ ^\cdot compound \qquad (33)$$

The rate of formation of RI would depend on the specific radical and iodine compound involved. The rate of H atom abstraction by OH or I radicals is fast; it is much slower for larger radicals. The rate of reaction of organic radicals with I_2 is rapid, whereas the rate of reaction of other iodine species is expected to be much slower.

Once formed, either by chemical reactions or by radiation-induced reactions, organic iodides will also be subject to radiolysis. Although the radiolytic reactions are somewhat specific to the organic iodide species involved, methyl iodide and other alkyl halides react rapidly and efficiently with e^- and with H in aqueous solution. The major products are hydrogen peroxide, iodide, iodine, alkyl hydroperoxide and formaldehyde. The exact proportions depend on the pH; mostly I^- is formed at pH values above neutral, whereas a mixture of I^- and I_2 results in acid solutions.

Radiation can also have an indirect effect on the chemistry of iodine. It is known that, in a radiation field, N_2 is partly transformed into N_2^+ and N^+ ions, which can react with H_2O vapour and O_2, leading eventually to nitric acid, HNO_3 (25). This could lower the pH of the aqueous phase and thus increase iodine volatility. As discussed earlier, I_2 is thermodynamically more favoured at low pH, the rate of oxidation of I^- is faster at low pH, and the radiation-induced steady state leans more toward I_2 in acid media.

However, radiolytic formation of HNO_3 is suppressed in the presence of hydrogen (25). In the Three Mile Island-2 case, not only was the possible formation of HNO_3 inhibited by hydrogen, but its possible effect on the pH was counterbalanced by the injection of sodium hydroxide in the primary coolant, by the spray system containing sodium hydroxide, and by the release of about 100 kg of cesium hydroxide from the core.

Conclusion

An examination of the chemistry of iodine in nuclear fuel, in the reactor coolant system, and in the containment building leads to the conclusion that the chemical conditions at Three Mile Island-2 favoured low iodine volatility. The chemistry of iodine in nuclear fuel is such that iodine was almost certainly released to the primary coolant as CsI. This is supported by experimental work, as well as by thermodynamic and kinetic analyses of the system. The conditions in the primary coolant system were such that iodine would have remained as CsI and been discharged to the containment building as an iodide solution.

Thermodynamic considerations show that, in the containment building, equilibrium would favour low volatility. A kinetic analysis indicates that the behaviour of the system is controlled by kinetic factors, if iodide is the initial chemical form. The direct oxidation of iodide by dissolved oxygen is then the rate-determining step in the formation of volatile iodine species. This oxidation is very slow under the chemical conditions that were present in the containment building at Three Mile Island-2.

Although the low volatility of iodine observed at Three Mile Island was undoubtedly due to a variety of physical and chemical factors, such as plating on internal reactor components and on epoxy-lined concrete, settling on surfaces as well as precipitation out of solution, the aqueous chemistry of the iodine system played a major role.

Acknowledgments

The authors thank Dr. F. Garisto of the Whiteshell Nuclear Research Establishment, who performed the calculations illustrated in Figures 1 and 2. We also thank Ontario Hydro for financial support of this research through the CANDEV agreement. This work was issued as AECL-8743.

Literature Cited

1. "Reactor Safety Study", Nuclear Regulatory Commission, Study Director N.C. Rasmussen, WASH-1400, 1975.
2. "Technical Bases for Estimating Fission Product Behaviour During LWR Accidents", Nuclear Regulatory Commission, NUREG-0772, 1981.
3. Wilson, R. "Radionuclide Release from Severe Accidents at Nuclear Power Plants", to be published in Rev. Mod. Phys.
4. Besmann, T.M.; Lindemer, T.B. Nucl. Tech. 1978, 40, 297.
5. Cubicciotti, D.; Sanecki, J.E. J. Nucl. Mat. 1978, 78, 96.
6. Lorentz, R.A; Collins, R.A.; Malinauskas, A.P.; Kirkland, O.L.;

Towns, R.L. "Fission Product Release from Highly Irradiated
Fuel", Nuclear Regulatory Commission, NUREG/CR-0722
(ORNL/NUREG/TM-287), 1980.

7. Kleykamp, H. "Formation of Phases and Distribution of Fission
    Products in an Oxide Fuel", Proc. Behaviour and Chemical State
    of Irradiated Ceramic Fuels, 1972, p. 157.

8. Lorentz, R.A.; Collins, J.L.; Malinauskas, A.P. "Fission
    Product Source Terms for the LWR Loss-of-Coolant Accident",
    Nuclear Regulatory Commission, NUREG/CR-1288
    (ORNL/NUREG/TM-321), 1980.

9. Ardron, K.H.; Cain, D.G. "TMI-2 Accident Core Heat-Up
    Analysis", Nuclear Safety Analysis Center, NSAC-24, 1981.

10. Garisto, F. "Thermodynamics of Iodine, Cesium and Tellurium in
    the Primary Heat-Transport System Under Accident Conditions",
    Atomic Energy of Canada Limited Report, AECL-7782, 1982.

11. Potter, P.E.; Rand, M.H. Calphad 1983, 7, 165.

12. Hahn, R.; Ache, H.J. Nucl. Tech. 1984, 67, 407.

13. Sallach, R.A. "Chemistry of Fission Product Elements in High
    Temperature Steam. I. Thermodynamic Calculations of Vapor
    Composition", Sandia National Laboratories, SAND81-0534/1,
    1981.

14. Elrick, R.M.; Sallach, R.A. "Fission Product Chemistry in the
    Primary System", Proc. Int. Meet. on LWR Severe Accident
    Evaluation, 1983.

15. Pelletier, C.A.; Thomas, C.D.; Ritzman, R.L.; Tooper, F.
    "Iodine-131 Behavior During the TMI-2 Accident", Nuclear Safety
    Analysis Center, NSAC-30, 1981.

16. Lemire, R.J.; Paquette, J.; Torgerson, D.F.; Wren, D.J.;
    Fletcher, J.W. "Assessment of Iodine Behaviour in Reactor
    Containment Buildings from a Chemical Perspective", Atomic
    Energy of Canada Limited Report, AECL-6812, 1981.

17. Paquette, J.; Lemire, R.J. Nucl. Sci. Eng. 1981, 79, 26.

18. Wren, D.J.; Sanipelli, G.G. "The Volatility of HOI", Proc. ANS
    Topical Meeting on Fission Product Behaviour and Source Term
    Research, 1984.

19. Toth, L.M.; Pannell, K.D.; Kirkland, O.L. "The Aqueous
    Chemistry of Iodine", Proc. ANS Topical Meeting on Fission
    Product Behaviour and Source Term Research, 1984.

20. Sigalla, J.; Herbo, C. J. Chim. Phys. 1957, 54, 733.

21. Burns, W.G.; Marsh, W.R. "The Decomposition of Aqueous Iodide
    Solution Induced by γ-radiolysis and Exposure to Temperatures
    up to 300°C", Proc. Int. Conf. Water Chemistry of Nuclear
    Reactor Systems 3, Vol. 1, 1983, p. 89.

22. Paquette, J.; Ford, B.L.; Wren, J.C. "Iodine Aqueous Chemistry
    Under Reactor Accident Conditions", Proc. ANS Topical Meeting
    on Fission Product Behaviour and Source Term Research, 1984.

23. Paquette, J.; Torgerson, D.F.; Wren, J.C.; Wren, D.J. J. Nucl.
    Mat. 1985, 130, 129.

24. Lin, C.-C. J. Inorg. Nucl. Chem. 1980, 42, 1101.

25. Linacre, J.K.; Marsh, W.R. "The Radiation Chemistry of
    Heterogeneous and Homogeneous Nitrogen and Water System",
    Harwell, AERE-R10027, 1981.

RECEIVED August 6, 1985

THE CLEANUP

11

Development of the Flowsheet Used
for Decontaminating High-Activity-Level Water

E. D. Collins[1], D. O. Campbell[1], L. J. King[1], J. B. Knauer[1], and R. M. Wallace[2]

[1]Oak Ridge National Laboratory, Oak Ridge, TN 37831
[2]Savannah River Laboratory, Aiken, SC 29801

Samples of actual high-activity-level water from TMI-2
were used to develop a chemical processing flowsheet
for decontamination of the water and concentration of
the radioactive contaminants in a form suitable for
disposal. The process included (1) sorption of the
bulk radioactive materials, cesium and strontium, onto
an inorganic ion exchanger; and (2) sorption of the
remaining traces of cesium and strontium plus anionic
contaminants, such as antimony and ruthenium, onto
standard organic ion exchangers. The latter step was
accomplished by means of a special deionization/
sorption technique. Prior to its use, the initial
step (removal of the bulk radioactive materials) was
improved by evaluating mixtures of zeolite ion
exchangers and selecting a combination which enabled a
significantly greater volume of water to be processed
through the exchanger.

The accident at Three Mile Island Nuclear Power Station, Unit 2
(TMI-2) resulted in the generation of ~2800 m³ of high-activity-
level contaminated water. In perspective, this water contained
~100 times the amount of radioactive contaminants as that generated
annually in radwastes from all nuclear power stations in the United
States. However, this amount was still several orders of magnitude
lower than would be typical in the wastes from a fuel reprocessing
plant.
    The objective of the work described herein was to develop a
process for decontaminating the high-activity-level water and for
concentrating the radioactive contaminants in a form suitable for
disposal. Decontamination of the water to concentration levels
specified by 10 CFR 20, Appendix B, Table II, Column 2 for release
to the environment was desirable even though specially imposed
restrictions on TMI-2 water prevented its release. The decontamination goal required that the concentrations of radioactive contaminants be reduced by factors as high as 8×10^7.
    Soon after the accident, samples of the reactor coolant system
(RCS) water were sent to the Oak Ridge National Laboratory (ORNL)

0097–6156/86/0293–0212$06.00/0
© 1986 American Chemical Society

for analyses of the chemical and radiochemical constituents. The samples were also used to evaluate potential methods for decontaminating the high-activity-level water and concentrating the radioactive contaminants into a readily disposable form.

Based on the analyses and test results, potential decontamination processes were considered and recommendations were made to the TMI-2 Technical Advisory Group (TAG). The TAG selected a process which was based primarily on sorption of the bulk radioactive components, cesium and strontium, onto an inorganic ion exchanger, Linde Ionsiv IE-96. This exchanger was the sodium form of a chabazite-type of zeolite that was commercially available and had a history of successful, large-scale usage. Standard organic ion exchange resins were to be used to sorb the remaining traces of radioactive contaminants.

The processing system was designed by Allied-General Nuclear Services for Chem-Nuclear Systems, Inc., the prime contractor for equipment fabrication and installation. The process was designed so that the equipment items that were to contain high levels of activity were housed in one of the spent fuel handling pools in order to use the pool water for shielding. Therefore, the process was called the "Submerged Demineralizer System," or SDS — although the process was not intended to demineralize the water during its decontamination.

The original SDS flowsheet was evaluated in a series of tests made at ORNL using 3 L of TMI-2 Reactor Building sump water (1). The results showed that the bulk of the cesium and strontium was sorbed on the zeolite, as expected, but the subsequent treatment with organic-based polishing resins would not provide additional decontamination from cesium and strontium or removal of the minor contaminants, ^{125}Sb and ^{106}Ru. Therefore, it became necessary to conduct process improvement tests (2,3).

The initial step of the flowsheet (i.e., removal of the bulk contaminants) was improved by evaluating mixtures of zeolite ion exchangers and selecting a combination which enabled a significantly greater volume of water to be processed through the exchanger. During that work, a mathematical model was developed to predict the performance of the SDS process. Also, a special deionization/sorption technique was developed to remove the remaining radioactive contaminants from the zeolite effluent.

Composition of the High-Activity-Level Water

The most important chemical and radiochemical components, the concentrations in each body of water which existed at the time of process development, and the total amounts present are listed in Table I. The high-activity-level water consisted of two bodies — ~340 m^3 that remained in the closed-loop recirculating reactor coolant system and ~2440 m^3 that had spilled onto the reactor containment building floor. The first samples of RCS water were obtained within a few days after the accident, and subsequent samples were taken periodically. However, the larger volume on the containment building floor could not be sampled until an access probe was installed about 5 months after the accident.

Table I. Composition of High-Activity-Level TMI Water
(Values are corrected for radioactive decay to July 1, 1980.)

Parameter	Reactor coolant system (RCS)	Containment Building water (CBW)	Total
Volume (m³)	340	2440	2780
Concentration (mg/L)			
Sodium	1350	1200	3400 kg
Boron	3870	2000	35,000 kg[a]
Cesium	1.5	0.8	2.5 kg
Strontium	<0.05	0.1	0.3 kg

	RCS		CBW		
Nuclide	Conc. (μCi/mL)	Relative ingestion hazard[b]	Conc. (μCi/mL)	Relative ingestion hazard[b]	Total (Ci)
^{3}H	0.17	60	1.0	300	2,500
^{89}Sr	5[c]	2,000,000	0.53	200,000	3,000
^{90}Sr	25[c]	80,000,000	2.3	8,000,000	14,000
^{106}Ru	0.1	10,000	0.002	200	40
^{125}Sb	0.01	100	0.02	200	50
^{134}Cs	10	1,000,000	26	3,000,000	67,000
^{137}Cs	57	3,000,000	160	8,000,000	410,000
^{144}Ce	0.03	2,000	0.0005	50	10

[a]As boric acid.
[b]Expressed as multiples of the concentrations listed in 10 CFR 20, Appendix B, Table II, Column 2.
[c]Values vary, probably because of precipitation.

Both bodies of water contained primarily sodium borate and boric acid, and the pH values were 8.2 and 8.6, respectively. The boron content of all of the water represented a total of 35 tons of boric acid. This fact was particularly significant when considering evaporation or total demineralization as a method for concentrating the radioactive contaminants and decontaminating the bulk of the water.

Although sodium and boron were the key chemical contaminants, it is significant to note that cesium and strontium were the primary radioactive materials, and the radiocesium isotopes in both bodies of contaminated water were, by far, the predominant sources of gamma activity. These isotopes necessitated that the decontamination process equipment be shielded and operated remotely to prevent excessive exposure to operating personnel. The strontium concentration was somewhat lower than that of cesium, in terms of radioactivity, but both were equally hazardous to human ingestion.

In addition to cesium and strontium, one of the radioactive contaminants present in the waters was tritium. This heavy isotope of hydrogen could not be removed by any practical separations process; however, the concentration of tritium in the high-activity-level water was sufficiently small so that removal of tritium from the water was not necessary.

In addition to the water-soluble contaminants, a significant concentration of strontium was found in an insoluble form in samples of water taken from the bottom of the Containment Building. In each sample, the concentration of solids in the slurry (liquid plus solids) was about 0.5% by volume, as determined by centrifugation; however, both the amount and the nature of the solid material in the slurry sample may not have been representative of the total solids within the building since the sample was taken from only one location. The key chemical and radiochemical constituents in the solids are listed in Table II. Also, Table II shows the calculated percentage of each element and nuclide in the total sample (liquid plus solid) that was in the solid phase. This calculation is based on the composition of 99.5% liquid, 0.5% solid in the sample, and on the concentrations of chemical elements and radioactive nuclides in the two phases. (The concentrations in the solid phase are shown in Table II, but those in the liquid are not.)

Process Flowsheets Considered for Decontamination of the Water

All the processes considered for the decontamination included ion exchange or evaporation, or both, as indicated by the flowsheets presented in Figure 1.

The process flowsheets were compared on the bases of the estimated volume of waste concentrates generated and the potential operating and maintenance problems. In general, the flowsheets that would generate the smaller volumes of waste were those which would allow selective removal of the radioactive contaminants while leaving the boron and sodium in the water.

The first flowsheet option (Figure 1) is a conventional ion exchange process such as that used for decontaminating low-activity-level "radwaste" water during normal operation at nuclear power stations. This method is not satisfactory for sorbing large amounts

Table II. Analysis of Solids in the Containment Building Water

Chemical			Radiochemical		
Chemical element	Conc. (mg/L solids)	Percent in solid phase[a]	Radioactive nuclide	Conc. (μCi/mL solids)[b]	Percent in solid phase[a]
Copper	7500	99	^{90}Sr	38	8
Nickel	2500	>98	^{89}Sr	8.7	8
Aluminum	1450	88	^{137}Cs	4.7	0.04
Iron	850	81	^{125}Sb	1.5	28
Silicon	650	10	^{144}Ce	1.4	93
Calcium	450	7	^{134}Cs	0.82	0.04
Zinc	400	>87	^{106}Ru	0.76	66
Chloride	400	10	^{95}Nb	0.14	97
Magnesium	150	10	^{60}Co	0.073	88
Sulfur	100	c	^{103}Ru	0.010	66

[a]Percentage of element or nuclide in total sample (liquid plus solid) that is in the solid phase. Calculation based on solids content of 0.5 vol % in samples, as determined by centrifugation.
[b]Concentration on July 1, 1980.
[c]Not measured in water.

of radioactive material because the resulting radiolytic degradation of the organic resins would interfere with subsequent resin solidification, transfer, and storage operations.

In the second flowsheet, periodic removal of cesium and strontium from the cation exchange resin via elution with acid would minimize the long-term radiation exposure and enable a smaller volume of resin to be used. The acid solution, containing most of the sodium originally in the water as well as the highly radioactive isotopes, would be concentrated by evaporation. The minimum volume obtained would be limited by the sodium sulfate concentration, which can be increased to about 22% in typical operations. However, experience at nuclear power stations had indicated that evaporators required frequent maintenance, and this would be significantly more difficult since the maintenance of an evaporator containing highly radioactive materials would have to be done remotely.

The maintenance problems would be even more severe if direct evaporation was attempted, as illustrated in the third flowsheet. In addition, a larger volume of high-level waste would be generated because the concentrate would contain the 35 tons of boric acid.

The fourth flowsheet would utilize an inorganic ion exchanger for sorption of most of the highly radioactive cesium and strontium, followed by evaporation for removal of all remaining radionuclides, except tritium. Inorganic ion exchangers, such as zeolites, are known to have a much greater degree of radiation stability than organic-based resins and a very high selectivity for cesium. Large-scale, successful operations using zeolites are a matter of record at several installations. In this case, the evaporator to be used would not contain the highly radioactive material; therefore, its operation and maintenance would not be as difficult as in the second or third option. Another advantage of this flowsheet would be that the use of evaporation as a polishing step would provide dependable

decontamination of the water as compared with ion exchange processes, which could be ineffective if nonionic species and colloids
were present. The only disadvantage of this scheme would be that
the evaporation would produce a large volume of boric acid concentrate, even though the concentrate would be low-level waste in
this instance.

The fifth flowsheet would also use an inorganic ion exchanger
for removing most of the cesium and strontium but would use organic-
based resins for polishing. The latter step would be effective for
decontaminating the water if nonionic species and colloids were not
present in significant concentrations. Overall, this last process
would generate the lowest volume of waste concentrates and, thus,
it was selected for use at TMI-2.

Evaluation of Potential Sorbents

Because of the need to expedite process design and equipment fabrication, a rapid process selection was made on the basis of the
considerations described above and the results of a few tests made
with the small samples of RCS water that were available.

Distribution coefficients between the RCS water and selected
sorbents were measured, and small scale ion exchange column tests
were made using synthetic solutions traced with radioactive cesium
or strontium. These trials were made to compare the loading performance of the various sorbents of interest and to determine the
effect of some of the process variables.

The breakthrough curves in Figure 2 show the superiority of
zeolites for cesium sorption. IE-95 is a chabazite type of zeolite
in the calcium form; however, the sodium form (called IE-96) was
selected for use at TMI-2.

Sodium titanate was not a good sorbent for cesium, but it
was the best one found for strontium. Unfortunately, the titanate
had only been produced in experimental amounts, and it had a soft,
powdery texture which was not suitable for use in large-scale
columns. It was tested as a mixture with IE-95, as shown in
Figure 3, but was not considered further.

The kinetics of strontium sorption were generally slower than
for cesium, and longer column residence time was required. If the
column residence time was at least 10 min, IE-96 zeolite was found
to be an acceptable sorbent for strontium. Thus, IE-96 zeolite was
selected by the TAG as the best sorbent for both cesium and strontium.

Evaluation of the Original Flowsheet

The original SDS flowsheet, shown in Figure 4, provided for the contaminated water to be clarified and then passed through a series of
ion exchange columns. Four small columns, each containing about 225
L of sorbent, were to be located within the spent fuel pool. The
first three were to contain IE-96 zeolite, while the fourth was to
contain a strong-acid type cation exchange resin. The manner of
operation was designed to accommodate the needed contact time (>10
min) for strontium sorption.

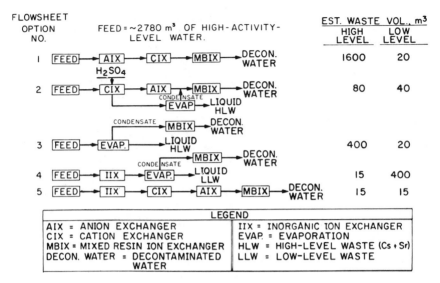

Figure 1.   Process flowsheets considered for decontamination of high-activity-level water.

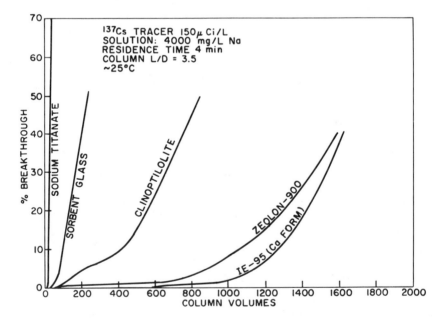

Figure 2.   Effectiveness of several inorganic sorbents for cesium loading.

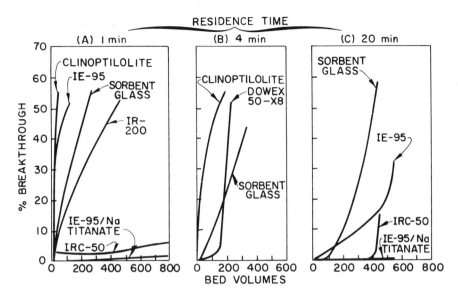

Figure 3. The effect of residence time on strontium loadings for a variety of sorbents.

^{89}Sr tracer (150 $\mu Ci/L$ — 4000 mg/L Na — 25°C).

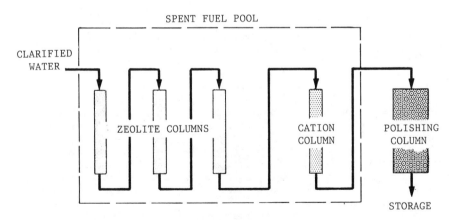

Figure 4. Original SDS flowsheet. "Reproduced with permission from Ref. 3. Copyright 1982, American Institute of Chemical Engineers."

The columns were modular and were intended to be used as the
radioactive waste containers after being loaded. The flowsheet
called for the columns to be moved after each had processed 50 m^3 of
water. This volume was equivalent to about 200 bed volumes, based
on each column. At that point, the column in the first position
would be discharged, the other two moved forward one position countercurrent to the flow of water, and a new column installed in the
third position. In this manner, the cesium would be loaded onto the
column in the first position and flow through all three columns
would provide sufficient contact time for strontium sorption.

This was a conservative design and would have required at least
60 columns to process all of the high-activity-level water.

The SDS process was evaluated in a series of small-scale tests
using 3 L of TMI-2 Containment Building water. Results showed that
the bulk of the cesium and strontium was effectively adsorbed on the
zeolite; however, the subsequent treatment with organic-based cation
and polishing resins did not provide additional decontamination.

Process Flowsheet Improvements

The objectives of the process improvement tests were (1) to obtain
increased loadings of cesium and strontium onto the zeolite columns
and (2) to develop an effective method for the polishing decontamination of the effluent water from the zeolite columns.

Improved Loading of Zeolite Columns. A 1000-bed volume zeolite test
had been made during evaluation of the SDS flowsheet, and conclusions were that as many as about 600 bed volumes could be processed
before strontium breakthrough from the third column would occur.
Since the organic cation resin (which originally was to be used in
the fourth column) was found to be ineffective for providing additional decontamination, consideration was given to using zeolite in
the fourth column in order to provide backup capability. Then, the
throughput could be increased to at least 600 bed volumes while
maintaining a sufficient safety margin for strontium.

A further increase in loading was envisioned after a more
"strontium-specific" zeolite, Linde A-51, was identified. Subsequently, the use of IE-96 zeolite for cesium sorption and A-51
zeolite for strontium sorption in either mixed beds, layered beds,
or alternate columns was considered. The use of alternate columns
would require that two types of columns would have to be maintained
and perhaps treated differently during subsequent waste solidification operations. The use of layered columns might mean that the
bottom layer would be adversely affected if flow distribution at the
bottom of the column was inefficient. Therefore, the use of mixed
beds appeared to be the most reasonable approach.

A 1500-bed volume test was then made with TMI-2 Containment
Building water and a mixed zeolite containing equal parts of IE-96
and A-51. Figure 5 compares the breakthrough curves obtained in
that test for cesium and strontium with those obtained while using
only IE-96.

When only IE-96 zeolite was used, less than 0.01% cesium breakthrough occurred during the entire test, but the strontium broke
through early. When the mixed zeolite was used, the capacity for

strontium sorption was increased by a factor of about 10 while that for cesium was still adequate. Thus, a tenfold increase in column throughput was made possible.

A series of tracer-level experiments was made to determine the effect of the mixed zeolite ratio on cesium and strontium break-through. Since a sufficient volume of TMI-2 water was not available for these tests, a synthetic solution was formulated with a chemical composition similar to that of the TMI-2 Containment Building water and was traced with radioactive cesium and strontium.

Column beds with IE-96/A-51 ratios of 3/1, 2/1, and 1/1 were evaluated. A balanced loading of cesium and strontium was the goal; however, if any uncertainty existed, the choice was to obtain better cesium loading because breakthrough of the gamma emitters (radio-cesium isotopes) could not be tolerated. The results of the single-column tests, shown in Figure 6, indicated that the proper ratio for balanced loading of cesium and strontium was between 2/1 and 1/1 at the desired throughput of ~2000 bed volumes.

Modeling of Zeolite Column Performance. The tests provided break-through data for only one column. Therefore, the data were fitted by means of the mathematical "J-function," using the constant separation factor model developed by Thomas (4), to enable calcula-tion of the mass transfer coefficients and extrapolation of the data to obtain the estimated performance of a second, third, and fourth column in series.

The general (Thomas) equation for the reaction kinetics of ion exchange in a fixed bed is as follows:

$$- \left(\frac{\partial X}{\partial N} \right)_{NT} = \left(\frac{\partial Y}{\partial NT} \right)_{N} = X(1 - Y) - RY(1 - X) \tag{1}$$

where X and Y are the dimensionless concentrations of the solute ion in the fluid and solid phases, respectively, and R is the separation factor. The variable X is defined as C/C_0, where C and C_0 are the concentrations of the solute ion of interest in the effluent and feed solutions, respectively. The variable Y is defined as q/q^*, where q is the actual concentration in the solid phase, and q^* is the concentration in the solid phase when it is in equilibrium with fluid at the inlet concentration, C_0. When the concentration of the solute ion is small relative to the concentration of the replaceable ion in the feed (as it is in this case), R approaches unity and the isotherm is linear.

The variable N represents the length of the exchange column in transfer units and is defined by the expression

$$N = K_d' \rho_B K_a / (f/v) \tag{2}$$

in which K_d' is the distribution coefficient when X = 1, ρ_B is the bulk density of the ion exchanger, K_a is the mass-transfer coeffi-cient characteristic of the system, f is the rate of flow of solu-tion through the column, and v denotes the overall volume of the sorbent bed, including the void spaces. The throughput parameter, T, is defined approximately by:

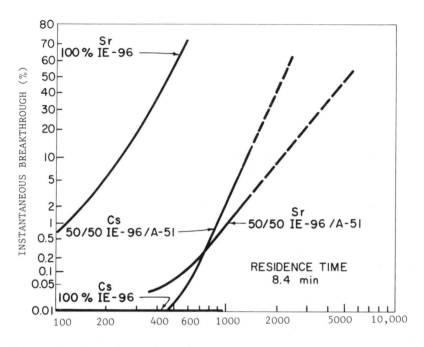

Figure 5. Comparison of results from single-zeolite and mixed-zeolite tests. Reproduced with permission from Ref. 3. Copyright 1982, American Institute of Chemical Engineers.

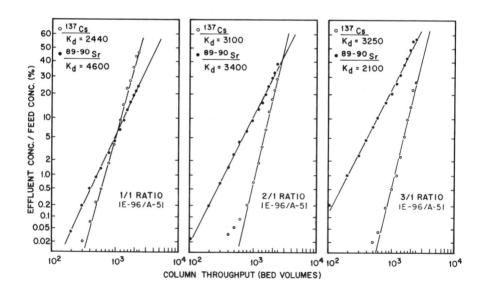

Figure 6. Results of tracer-level column tests.

$$T = (V/v)/K_d'\rho_B \qquad (3)$$

where V is the volume of solution processed through the column. Note that V/v is the number of "bed volumes" of solution. Since ρ_B is essentially constant, it is convenient to define a volume-basis distribution coefficient, $K_d = q_v/C_0$, where q_v is the concentration of the solute ion per unit volume of the sorbent bed (sorbent plus void space) and C_0 is the concentration in the feed solution. Equations (2) and (3) can then be expressed as

$$N = K_d K_a/(f/v) \qquad (2a)$$

and

$$T = (V/v)/K_d \qquad (3a)$$

Equation (1) has been integrated [Eq. (16-128a) in ref. 5] for the special case of reversible second-order reaction kinetics (appropriate to ion exchange), and the solution was found to be:

$$X = C/C_0 = \frac{J(RN,NT)}{J(RN,NT) + [1 - J(N,RNT)]\ \exp[(R - 1)N(T - 1)]} \qquad (4)$$

where J is a mathematical function (5) related to the Bessel function, I_0.

For large values of RN (a condition approached in SDS operation and in the small-scale tests), $C/C_0 = \sim 0.5$ when T = 1, independent of the values of RN. This characteristic can be employed in the data analysis. Experimental data can be used to construct logarithmic-probability plots of C/C_0 vs V/v. These plots are nearly linear and can be used to estimate K_d, which is approximately equal to V/v at the point where $C/C_0 = 0.5$ (at 50% in Figure 6). Values of R and N can then be obtained from the experimental data through the iterative use of Eqs. (3a) and (4).

A numerical-solution model of the Thomas equation was developed to accept input data in a form that simulates the cyclical mode of operation proposed for the SDS. A numerical solution was required to analyze the multibed system in which the partially loaded columns are moved forward (countercurrent to the water flow) periodically, because an analytic solution is not practical unless the initial loading on each bed is zero.

The four columns of the SDS were represented in the model by two 4000-point arrays (one each for X and for Y), using 1000 points for each column. Calculations were carried out to simulate the passage of the desired volume of feed through the four columns in series, with initial values of zero for X(n) and Y(n) for all points.

At the end of the first feed cycle, the values of X(n) and Y(n) were replaced by the previously calculated values of X(n + 1000) and Y(n + 1000) for values of n between 1 and 3000 and were set equal to zero for values of n between 3001 and 4000. This procedure simulated removing the first column, moving the last three columns forward one position, and putting a new column in the fourth position.

The calculations were then repeated for another cycle, using this configuration as the initial condition. This modeled rotation procedure was repeated for the number of cycles necessary to process the total volume of high-activity-level contaminated water.

Predicted and Actual Zeolite Column Performance. By interpolating between the experimentally derived distribution coefficients and the calculated mass transfer coefficients and separation factors for 1:1 and 2:1 zeolite mixtures, values for these parameters were derived for the 3:2 mixture. These values are shown as "test-column data" in Table III, along with corrected values obtained from early SDS operations. Both the mass transfer and the distribution coefficients were higher than predicted by the "test-column data" for cesium. For strontium, the distribution coefficient was lower but the mass transfer coefficient was higher than predicted by the "test-column data"; this situation resulted in a loading behavior similar to that predicted. Considering that the scale of the test column was 10^5 times smaller than the SDS columns, the comparison with performance data was considered to be excellent.

Table III. Comparison of Test-Column and SDS-Column
Performance Parameters for Cesium and Strontium

Parameter	Test-column data		SDS-column data	
	Cesium	Strontium	Cesium	Strontium
Distribution coefficient, K_d	2805	3760	3800	3000
Mass transfer coefficient, K_a	8.0×10^{-4}	2.9×10^{-4}	2.2×10^{-3}	5.6×10^{-3}
Separation factor, R	1.15	1.65	1.0	1.0

Using the early SDS data, cesium and strontium breakthroughs were calculated for six loading cycles in which 2760 m^3 of high-activity-level contaminated water (HALW) was processed (460 m^3 in each cycle). The results are shown in Table IV. Although the strontium breakthrough continued to increase throughout the six cycles, the breakthrough from the fourth column did not exceed the concentration (0.1%) of the nonexchangeable species of strontium that had been experimentally observed in the HALW. Based on these calculations, the number of zeolite columns needed to process the bulk of the HALW at TMI-2 was reduced to ~10.

Improvement of Polishing Decontamination Method. Results from the evaluation tests on the original SDS flowsheet showed that the effluent water from the zeolite columns contained residual cesium and strontium at concentrations of about 10^{-3} μCi/mL each. Also, the effluent water contained anionic nuclides, [125]Sb and [106]Ru, at concentrations of 10^{-2} and 10^{-3} μCi/mL, respectively.

Filtration tests indicated that the residual cesium and strontium were either in the form of nonionic colloids or adsorbed on colloids of other materials. Thus, these contaminants could not be removed by ion exchange methods unless the water was treated to change the chemical nature of the residual cesium and strontium.

Table IV. Calculated Cumulative Breakthrough of Cesium and
Strontium for Six Cycles of Column Replacement

	Cumulative breakthrough (% of feed)							
	Cesium				Strontium			
Cycle	Column 1	Column 2	Column 3	Column 4	Column 1	Column 2	Column 3	Column 4
1	0.62	a	a	a	13.8	0.18	a	a
2	0.71	a	a	a	22.9	0.92	a	a
3	0.73	a	a	a	29.5	2.14	a	a
4	0.73	a	a	a	34.5	3.62	0.17	a
5	0.73	a	a	a	38.3	5.21	0.35	a
6	0.73	a	a	a	41.4	6.81	0.60	a

[a]Calculated breakthrough less than observed concentrations (0.003% of cesium and 0.1% of strontium) of nonexchangeable species.

During the flowsheet evaluation tests, indications were that, if the zeolite effluent water could be allowed to age for at least several hours before contact with the polishing sorbents, further decontamination could be obtained. The theory was that the residual species were ionic but were sorbed on colloids of other materials in the water and that, if the exchangeable species were first removed from the water, the material sorbed on the colloids would reequili- brate with the water during the aging period and become susceptible to removal by subsequent ion exchange treatment.

A series of tests was designed to investigate the effects of aging times from 3.6 to 605 ks (1 h to 7 d) at (1) ambient con- ditions, (2) an elevated temperature (75°C), (3) a reduced pH level (pH = 6), and (4) combinations of these conditions. Further, a com- parison was made of the use of either IE-96 zeolite or a cation exchange resin (Nalcite HCRS, in the sodium form) for the polishing treatment after the aging period. The results of these tests showed that significant reductions of the cesium and strontium concentra- tions could be obtained by aging for at least 2 h at 75°C.

In contrast to the nature of the residual cesium and strontium, none of the antimony and only about one-third of the ruthenium were indicated to be in a colloidial form by ultra-high-speed centri- fuging tests. The antimony and ruthenium were initially sorbed in anion exchange column tests; however, breakthrough occurred early at a point which coincided with breakthrough of sodium from the pre- ceding cation exchange beds. Subsequently, tests were made to obtain an understanding of this effect.

In the tests, the water was pretreated by means of cation exchange with an acid-form resin. This treatment removed sodium ions and lowered the pH of the water. Several tests were made to lower the pH to different values. Then, in all of the tests, the pretreated water was treated further with a strong-base anion exchange resin (Nalcite SBR, in the borate form), and the equilib- rium distribution coefficients were measured. The results (Table V) showed that, as the pH was lowered, an exponential increase in the distribution coefficient occurred. These results can be explained by the hypothesis that the contaminants were not sorbed effectively by the anion exchange resin at the higher pH levels because of com- petition from the large concentration of borate ions in the water.

Table V.   Sorption of Residual Radioactive Components
by Strong-Base Anion Exchange Resin

pH[a]	Distribution coefficient with SBR ($H_2BO_3^-$)			
	^{106}Ru	^{125}Sb	^{90}Sr	^{137}Cs
7.8	16	29	1	1
7.3	54	210	1	5
6.4	2200	1500	21	23

[a]pH adjustment by means of cation exchange.

Further, by removing the sodium, the borate ions were converted to
weakly-ionized boric acid, thereby removing the competitive effect.
Thus, removal of the sodium was, in effect, a deionization of the
water.

The results also indicated that a smaller, but still significant,
reduction of the residual cesium and strontium concentrations could
be obtained at the lower pH levels.

The penalty for using this method would be the generation of a
relatively large volume of low-activity-level waste ion exchange
resin, which would be necessary to sorb the sodium. In comparison
to the high-activity-level zeolite wastes that would be generated
during removal of the cesium and strontium, a 20-fold greater volume
of cation and anion exchange resin would be required for removal of
the sodium and the anionic contaminants.

Other methods were also tried, including neutralization of the
sodium by addition of various acids, evaluation of other anion
exchange resins, use of a boron-complexing agent, and use of a
variety of other sorbents (e.g., high-surface-area glass, titanates,
zeolites, and molecular sieves). However, none of the methods
appeared to be usable except the one in which the pH is reduced by
means of sodium removal on a cation exchange resin followed by sorption of the contaminants on anion exchange resin.

Summary

A few small samples of high-activity-level water from TMI-2 were
used to develop a chemical processing flowsheet for decontamination
of the water and concentration of the radioactive contaminants in a
form suitable for disposal. The initially selected process was
evaluated and significantly improved. The improved process included
(1) sorption of the bulk radioactive materials, cesium and strontium, onto a mixture of inorganic zeolites; and (2) sorption of the
anionic contaminants, antimony and ruthenium, plus the remaining
traces of cesium and strontium onto standard organic ion exchange
resins. The latter step was accomplished by means of a special
deionization/sorption technique.

Literature Cited

1. Campbell, D. O., et al. "Evaluation of the Submerged Demin-
    eralizer System (SDS) Flowsheet for Decontamination of
    High-Activity-Level Water at Three Mile Island Unit 2 Nuclear
    Power Station," ORNL/TM-7448, 1980.

2. Campbell, D. O., et al. "Process Improvement Studies for the Submerged Demineralizer System (SDS) at the Three Mile Island Nuclear Power Station, Unit 2," ORNL/TM-7756, 1982.
3. Collins, E. D., et al. "Water Decontamination Process Improvement Tests and Considerations," AIChE Symp. Ser. 213, 1982, 78, 9.
4. Hiester, N. K.; Vermeulen, T.; Klein, G. "Chemical Engineers' Handbook," 4th ed. (Perry, J. H., et al., Eds.) McGraw-Hill, New York, 1963; p. 16-2.
5. Hiester, N. K.; Vermeulen, T. Chem. Eng. Prog., 1952, 48, 505.

RECEIVED July 16, 1985

12

Water Cleanup Systems

Cynthia G. Hitz and K. J. Hofstetter

GPU Nuclear Corporation, Middletown, PA 17057

Six years after the Three Mile Island
Unit 2 (TMI-2) accident, all of the
accident-generated water has been processed
and is being retained on site in the processed
water storage tanks (PWSTs). The two systems
used to process this water were the EPICOR II
system and the submerged demineralizer system
(SDS). Nearly 3 million gallons of water have
been processed by these two systems. The
EPICOR II system was used to process the
water that was originally located in the
auxiliary building. The SDS processed water
from the reactor building basement and the
reactor coolant system (RCS). SDS continues
to process water that has been used for decontamination. The systems are fundamentally
different in that the SDS uses zeolites to
remove radioactivity from the water and the
EPICOR system uses standard organic ion
exchange resins. The operational history
of these systems is discussed in this chapter.

A third water processing system is scheduled for operation during
the summer of 1985. This system, known as the defueling water
cleanup system (DWCS), will also use a zeolite to remove cesium
from the water in the refueling canal, spent fuel pool, and reactor
vessel during defueling. The DWCS will include an extensive
filtration operation to remove the suspended solids and fuel
particles that would result in turbidity.

The accident at Three Mile Island Unit 2 (TMI-2) Nuclear Generating
Station resulted in a release of significant quantities of
radioactive fission products from the reactor fuel to various parts
of the plant. In particular, large quantities of water containing
these radioactive contaminants were produced, which represented a
mobile medium by which the fission products would be dispersed.
Almost immediately following the accident, action commenced to

0097-6156/86/0293-0228$06.00/0
© 1986 American Chemical Society

design and install liquid radwaste processing systems that would immobilize these contaminants in solid waste form. Minimizing the exposure to operations personnel and the public were key design goals of these systems.

EPICOR II Radwaste System

System Development. The EPICOR system was designed and built to process the 565,000 gallons of radioactive waste water that was collected in the auxiliary building. Its entire conception and design development occurred during the four to six weeks following the TMI-2 accident. The auxiliary building water had been collected in many tanks and the chemistry and radiochemistry of these tanks varied considerably, as shown in Table I. A demineralizer system was selected for flexibility.

Table I. Auxiliary Building Water

Consituent	Units	Min	Avg	Max	
Boron	ppm	420	1073	2140	
Sodium	ppm	1	189	710	
Conductivity	mhos	695	1263	2800	
pH		7.68	8.18	9.2	
H-3	μCi/ml		0.0043		0.45
Cs-134		0.9		8.3	
Cs-137		4.5		46.0	
Gross βγ		7.3		67.0	

EPICOR II is located in the chemical cleaning building. This building, designed for steam generator cleaning, had not been used prior to the accident because of the excellent performance of the installed steam generators. This facility contained two large tanks, unused floor space, a sump, and a seismic bathtub foundation, making it ideal to house the EPICOR processing system. After four months of testing, operator training, and NRC review (including an Environmental Impact Statement and a ruling by the NRC Commissioners), the system became operational October 22, 1979.

System Description. The EPICOR II Radwaste System (Figure 5) is a demineralizer system comprising three (3) separate demineralizer beds. The resins are contained in liners which are fully portable

to enhance handling of spent resins containing high levels of fixed
radioactive contaminants. Two cylindrical liner sizes are used.
One, called a 4x4, is four feet in diameter by four feet high. The
other, a 6x6, is six feet in diameter by six feet high. Flow rate
through the system is 10 gallons-per-minute, provided by air-driven
sandpiper pumps. Each container is vented to the liquid effluent
tanks, which are vented atmospherically to the EPICOR II building
through filters. The EPICOR II building has an HVAC system for air
cleanup and to maintain the building at a negative pressure. This
enhances radioactive contamination control.

The great flexibility of EPICOR II permits use of either liner
size in all three positions. System configuration and operations
during the auxiliary building water processing are described below.

Prefilter/Demineralizer (First Liner). The first liner was a 4x4
containing resin material for cation removal. Organic and
inorganic ion-exchange were used to optimize the removal of
contaminants such as cesium (Cs), strontium (Sr), and sodium (Na).
Approximately 99% removal of these contaminants was achieved in
this liner. The prefilter was removed from service when it was
either chemically exhausted or had reached a maximum Curie (Ci)
content. To remain compatible with available licensed shipping
casks, this liner was limited to a maximum loading of 1,300
Curies. With a limited number of exceptions, this loading limit
was the predominant reason for liner changeout. The resin mixes
were optimized so as to achieve full removal of contaminants prior
to reaching the Curie limitations. A liner loaded with 1,300 Ci
resulted in radiation levels in excess of 1000R/hr. Fixed
shielding around and over the liner limited area dose rates to
5 mR/hr for easy access to the liner during service. The removal
of the liner for placement in the storage facility or shipping cask
required the development of a specially designed transfer bell.
The bell was equipped with retractable doors on the bottom which,
when closed, resulted in the liner being completely encompassed.

Middle Demineralizer (Second Liner). The second liner was also a
4x4 vessel containing resin material for cation removal. The
principal purpose of this liner was to provide additional cation
polishing for breakthrough from the first liner. This was
particularly critical during the end of a batch. This liner was
also designed to be removed from service at the 1,300 Curie
deposition limit (based on shipping limitations). However, because
of the high removal efficiency of the first liner, this second
liner never received more than 100 Curies deposition. The reasons
for changeout included chemistry control parameters such as sodium,
boron, and/or pH in the process stream.

Polishing Demineralizer (Third Liner). The third liner was a
cylindrical vessel six feet in diameter and six feet high. It was
the final polishing demineralizer and was used to remove anion

contaminants and reduce trace radioactive contaminants down to a concentration level on the order of 10^{-7} µCi/ml or less. This liner used mixed cation and anion resins to achieve this purpose. Antimony (Sb-125) had to be removed by the anion resin in this liner and, therefore, a careful balance between anion and cation resin ratios was required. Procedures limited this liner to 20 Curies of deposited activity because the transfer bell only accommodates the 4x4 liners. To date, no liner has contained more than 10 curies. Changeout was based on exhaustion and the ability to remove trace levels of activity form the process stream.

Tanks. Two tanks were provided in the system for monitoring and holdup purposes. Effluent was collected and sampled in one monitor tank. The other tank was used to store water requiring recycle through the system.

Instrumentation.   Instrumentation related to process controls includes:

o   In-line pH meters
o   In-line conductivity cells
o   In-line temperature thermometer
o   In-line liquid activity levels
o   Liquid flow rate
o   Liquid pressures
o   Liner radiation readings

To minimize personnel exposure, these instruments are monitored in a control room remote from the EPICOR II building. Other instrumentation is available to monitor radiation and airborne contamination levels, tank levels, and HVAC controls, to name a few.

System Performance. The system processed the 565,000 gallons of auxiliary building water between October 1979 and December 1980, removing over 56,000 Curies of Cesium from the water.
    A total of 72 liners were expended to achieve the auxiliary building cleanup. Not all liners were expended due to radiochemical limitations; some were removed from service due to level control adjustments. Typical liner throughputs achieved prior to liner changeout are:

Prefilter/Demineralizer:	5,000 to 12,000 gallons (Dependent upon radioisotopic concentrations of influent water.)
Middle Demineralizer:	80,000 gallons
Polishing Demineralizer:	150,000 gallons

SDS Radwaste System

System Development. The submerged demineralizer system (SDS) was designed to process the highly contaminated water in the reactor building basement (RBB) and reactor coolant system (RCS). An analysis of each of these waste streams has been provided in Table 2. To shield workers from the high levels of radioactivity in the waste streams, the system was located under water, hence its name. The high level of radionuclide concentration also dictated the use of inorganic zeolites. Inorganic zeolites were chosen because they are stable to ionizing radiation, are selective for the radionuclides and solutions containing competing cations such as sodium, and are compatible with the vitrification process. (Vitrification, in this context, refers to the U.S. Department of Energy's (DOE's) RD&D effort to encapsulate high-level waste in glass. DOE has agreed to accept TMI-2 wastes for use in this project.)

Table II presents the EPICOR performance for the highest activity water processed by the system.

Table II. EPICOR II Radwaste System Processing Performance
(µCi/ml)*

RADIO NUCLIDE	INFLUENT	EFFLUENT
CESIUM 134	7.0	7.54×10^{-6}
CESIUM 137	37.0	5.59×10^{-6}
STRONTIUM 89	0.4	2.79×10^{-6}
STRONTIUM 90	0.3	5.37×10^{-6}
TRITIUM	0.27	0.27

* – This reflects the highest activity water to be processed.

Many studies have shown organic ion exchange resins to be relatively unstable in high radiation environments. At a radiation exposure of approximately 10^8 Rad, organic ion exchangers begin to lose their functionality and generate hydrogen gas. Zeolites, on the other hand, have been exposed to radiation levels up to 10^{11} Rad with no loss of structure or functionality.

The high levels of sodium and boron in these two waste streams also required that the inorganic ion exchanger be highly selective for cesium and strontium over competing cations. Oak Ridge National Laboratory and Savannah River Laboratory conducted a series of column tests (reported in one of the other papers), which identified two suitable zeolites. The naturally occurring zeolite, chabazite, sold by Union Carbide under the trade name Ion Siv IE-96, was selected for cesium removal. A synthetic zeolite, also supplied by Union Carbide under the trade name Linde A-51, was very selective for strontium removal. These two were blended together, in the appropriate ratio, to remove the cesium and strontium

isotopes from the liquid waste streams. Since the zeolite
exchangers did not remove anions, some of the anionic radionuclides
(such as antimony) had to be removed downstream of the SDS. The
EPICOR system, which has so effectively cleaned up the water in the
auxiliary building, was used for this process. The ion exchange
system was reconfigured with two 6x6 liners in the first and second
position and a 4x4 liner in the third position. The functions of
each ion exchanger did not change even though the size did change
to accommodate the higher cation and anion concentrations in this
water.

System Description

Filtration and Staging Equipment. The SDS contains two filters,
known as the pre-filter and post-filter, used to filter influent
water prior to staging it to the tank farm. Two types of filters
have been used in the SDS. Initially, cuno filters were used.
These were mounted inside a ten cubic-foot vessel identical to the
ion exchangers. However, after several months of service, the NRC
became concerned about GPU Nuclear's ability to dewater this type
filter if it plugged with solids. Therefore, GPU Nuclear changed
to sand filters, which could be dewatered using the same procedure
as the ion exchangers.

The tank farm was a group of tanks, used in series, with a
combined volume of 60,000 gallons. These tanks were mounted in the
A spent fuel pool and were shielded by large concrete slabs. Water
was transferred to the SDS by a floating pump located in the RBB,
passed through the two filters, and was stored in the tank farm
tanks.

During SDS processing of the RCS, the reactor coolant was
staged to one of the installed reactor coolant bleed tanks (RCBTs)
and transferred to the SDS using the existing waste transfer
pumps. The reactor coolant passed through the filters and then
directly into the ion exchangers, bypassing the tank farm staging
tanks.

Ion Exchangers. The SDS consists of two parallel trains of four
ion exchange vessels. Each vessel has a capacity of 10 cubic feet
and is normally loaded with 8 cubic feet of the zeolite mixture.
Water is delivered to the ion exchangers from the tank farm via a
well type pump installed in the tank farm stand pipe. The normal
processing flowrate is five gallons-per-minute per train. The
effluent from the ion exchangers passes through a cartridge type
post-filter, which is designed to remove any zeolite fines
remaining in the process stream. The effluent is then collected in
two 11,000-gallon monitor tanks before being processed through the
EPICOR system.

Leakage Containment System. In order to keep spent fuel pool B
clean and free of contamination, the SDS also includes a subsystem
known as the leakage containment system. This system takes suction
from contamination control boxes around each of the ion exchangers
to ensure that process water escaping into the pool water from the
vessel connections is delivered directly to an ion exchange
system. Both organic and inorganic ion exchange materials have
been used in the leakage containment system. The zeolite material
performs better for cesium removal, while the organic has an edge
for strontium removal from the pool. An in-line radiation monitor
stops the process flow if a large increase in the pool
concentration is noted.

Other Support Systems. The SDS has several other support systems
that ensure its safe operation. A dewatering system, located in
another section of the B spent fuel pool, ensures that the SDS
liners are devoid of free-standing water prior to shipment to the
DOE for disposal. The SDS is also supplied with an offgas
ventilation system to ensure that any airborne contamination is
controlled and monitored and to permit hydrogen offgassing during
storage. Sample gloveboxes are provided to allow the operators to
safely sample even the most radioactive water processed through the
SDS.

System Performance. The SDS became operational in July 1981, first
processing accumulated auxiliary building water. This lower
specific activity water was chosen for the initial batches to
demonstrate system performance. The first three batches (150,000
gal) were processed through train 1, using only two ion exchangers
because of the lower activity. System performance was good, and
reactor building basement processing commenced following a small
test batch on train 2. By the time RBB processing started in
September 1981, the RBB depth had reached 8.5 feet. At the time of
the March 1979 accident, the majority of this water accumulated
from the RCS, the RB spray system, and a leak in the air cooler
coolant system. In the months following the accident, this
inventory continued to increase due to a small (approximately
0.1gpm) leak from the RCS. The 650,000 gallons of accumulated
water was processed over the next nine months. Normally, 44,000
gallons per batch was processed because this equated to four full
monitor tanks. The monitor tanks were then processed through the
EPICOR system and the effluent was stored in the PWST. After
110,000 gallons, the ion exchange vessel flowpath was
reconfigured. The first vessel was removed and the other three
vessels each moved forward one position. A new vessel was placed
in the fourth position.

Table III summarizes the SDS performance for RBB processing.

Table III. Average SDS/EPICOR Performance for Reactor Building
Sump Water Processing

RADIONUCLIDE	INFLUENT ($\mu Ci/ml$)	SDS EFFLUENT ($\mu Ci/ml$)	EPICOR EFFLUENT	SYSTEM DF*	CURIES REMOVED
Cs-134	13	1.0(-4)**	<2(-7)	>6.6(7)	29,800
Cs-137	123	8.6(-4)	3.2(-7)	3.8(8)	278,000
Sr-90	5.1	8.8(-3)	1.7(-5)	3(5)	11,600
Sb-125	1.1(-2)	1.1(-2)	<4(-7)	>2.7(4)	25
Ce-144	4(-4)	4(-4)	<1(-6)	>4(2)	1
Co-60	2(-5)	2(-5)	<2(-7)	>1(2)	0.05
H-3	8.8(-1)	8.8(-1)	8.8(-1)	1	0

*Decontamination Factor
**The number in parenthesis is the exponent [e.g., 1.0(-4) means
1.0×10^{-4}]

As shown on the table, the SDS removed 278,000 Ci of Cs-137,
29,800 Ci of Cs-134 and 11,600 Ci of Sr-90.

Following the highly successful RBB processing campaign, the
SDS was used to process the RCS water. Because the reactor core
must be covered with borated water, the RCS water was processed
using a feed and bleed method. Two RCBTs were used as the staging
tanks for this operation; the tank farm was bypassed. For each
batch, 50,000 gallons of RCS water was letdown to a RCBT while
50,000 gallons of low activity coolant-grade water was added,
maintaining the water level. The letdown RCS water was then
processed through the SDS for cesium and strontium removal. The
effluent was collected in another RCBT. This cycle was repeated
for each batch. The SDS was ideal for this processing because the
zeolites do not remove the sodium or boron, which are used to
control corrosion and core reactivity.

Five batches were processed by this method prior to the camera
inspection during the summer of 1982. Following the camera
inspection, numerous additional batches were processed until the
reactor vessel head was removed June 25, 1984.

Following head removal, a small pump was placed in the
internals indexing fixture (IIF). This pump was sized to transfer
the water directly from the IIF, through the sand filters and
through the SDS, to the RCBT. As with feed and bleed processing, a
RCBT of clean water was used to maintain the core liquid level.
This processing method will be used until the time of fuel removal,
currently scheduled for later in 1985.

The SDS has processed 815,000 gallons of RCS water. Processing
has reduced the cesium and strontium from their pre-processing
concentrations of about $14 \mu Ci/ml$ to less than 0.1 $\mu Ci/ml$ for

cesium, and 2.2µCi/ml for strontium. Because of an apparent
reappearance mechanism, continued processing is necessary to
maintain these lower levels.

In addition to processing water from the RBB and RCS, the SDS
has processed the water that was used for surface and gross
decontamination in the reactor building. This water, which was
previously processed through the SDS and EPICOR system, was
recycled for the decontamination operation. Periodically since the
SDS was started up in 1981, additional water that accumulated in
the auxiliary building sumps and tanks has also been processed
through the SDS. To date the SDS has processed a total of
2,700,000 gallons of waste water.

Defueling Water Cleanup System

System Development. The defueling water cleanup system (DWCS) has
been developed for use during the defueling effort to process the
water from the reactor vessel, the defueling canal, and spent fuel
pool A (SFPA). The current defueling plans maintain the water in
the reactor vessel at the current level in the IIF. The deep end
of the reactor canal will be isolated from the shallow end via a
dam. The deep end will be flooded to provide shielding over the
plenum and the fuel canisters during defueling; however, the
shallow end of the canal will not be filled. SFPA will be
connected hydraulically to the deep end of the canal through the
open fuel transfer mechanism tubes. The volumes of water in these
three locations are: 40,000 gallons in the reactor vessel; 70,000
gallons in the deep end of the canal; and 220,000 gallons of water
in SFPA.

The DWCS has two purposes. The first is to maintain water
clarity to allow the defueling operators to view tools and fuel
through the water. The second purpose of the DWCS is to remove the
soluble cesium from the water to minimize the dose to the operators
working above the reactor vessel.

The DWCS consists of two subsystems. One subsystem provides
cleanup for the reactor vessel, the other subsystem provides
cleanup for the deep end of the canal and SPFA. Each of the
subsystems provides filtration for clarity control and ion exchange
for radionuclide control.

System Description

Filtration Equipment. Camera examination of the core and samples
taken from the core region indicate that a large amount of very
small particles is present. Particles less than 5 microns in size
will tend to settle very slowly and, therefore, must be removed by
filtration.

Numerous requirements are placed on the filter system. Its primary purpose is to maintain water clarity; the system's effluent must be less than 1NTU. This value should permit clear visibility through 10-20 feet of water. Additionally, the material collected in the filter will be fuel debris, and therefore must be collected in a critically safe geometry. For this reason, the filter has been designed to fit inside the fuel canisters. This eliminates the need to qualify a second container as critically safe. Because the DOE has limited the number of canisters that will be accepted under the core contract, the total number of disposal containers must be minimized. Also, the total number of filters in service has to be minimized because of the limited amount of storage space in the deep end of the canal. Finally, the filter must be constructed of materials that have a high radiation stability, since the ultimate disposition of the TMI fuel canisters has not been determined by DOE.

A survey was made of the commercially available filters, and none were found to meet all of these filter requirements. Therefore, a new filter medium has been developed for use in this application. The new filter is constructed of 316L sintered stainless steel, pleated into a filter cartridge resembling a pleated paper filter cartridge. These filter cartridges are arranged in elements that are 11 filter units long. Seventeen elements then fit inside a filter canister. In this configuration, the filters meet all of the requirements that were set forth.

Each filter canister is rated at 100 gallons-per-minute flow. Four canisters are arranged in parallel in each subsystem, for a filtration flow rate of 400 gallons-per-minute. To determine the filter performance, an extensive test program was undertaken at the B&W Lynchburg Research Center. (This filter testing is the subject of one of the other papers.)

Ion Exchange. Based upon the successful operation of the SDS, zeolite ion exchange has been chosen for the DWCS. As mentioned previously, zeolite has the added advantage of not removing sodium and boron (used for corrosion and core reactivity control). The reactor vessel portion of the DWCS contains two 4x4 zeolite ion exchangers. These ion exchangers are identical to the 4x4s that are used in the EPICOR system, but they contain only zeolite material. With both ion exchangers operating, the processing flow rate is 60 gallons per minute. This flow rate can accommodate normal operations and upset conditions, which might be seen when the fuel is disturbed.

The deep end of the canal and SFPA are equipped with one 4x4 ion exchanger, since upset conditions are not anticipated in that body of water.

System Performance

The defueling water cleanup system is currently under construction
and scheduled for startup later this year, prior to the beginning
of defueling. Therefore, no performance data are currently
available. However, based on the filtration test work done at the
B&W Research Center and the proven performance of the zeolites in
the SDS, we anticipate excellent performance from the defueling
water cleanup system.

RECEIVED July 16, 1985

Defueling Filter Test

J. M. Storton and J. F. Kramer

Research & Development Division, Babcock & Wilcox, Lynchburg, VA 24506-1165

The Three Mile Island Unit 2 Reactor (TMI-2) has sustained core damage creating a significant quantity of fine debris, which can become suspended during the planned defueling operations, and will have to be constantly removed to maintain water clarity and minimize radiation exposure. To accomplish these objectives, a Defueling Water Cleanup System (DWCS) has been designed. One of the primary components in the DWCS is a custom designed filter canister using an all stainless steel filter medium. The full scale filter canister is designed to remove suspended solids from 800 microns to 0.5 microns in size. Filter cartridges are fabricated into an element cluster to provide for a flowrate of greater than 100 gals/min. Babcock & Wilcox (B&W) under contract to GPU Nuclear Corporation has evaluated two candidate DWCS filter concepts in a 1/100 scale proof-of-principle test program at B&W's Lynchburg Research Center. The filters were challenged with simulated solids suspensions of 1400 and 140 ppm in borated water (5000 ppm boron). Test data collected includes solids loading, effluent turbidity, and differential pressure trends versus time. From the proof-of-principle test results, a full-scale filter canister was generated.

The Three Mile Island Unit 2 Reactor (TMI-2) Recovery Operation presents a significant challenge to the nuclear industry. An important aspect in the TMI-2 Recovery effort is the water treatment techniques used to maintain water clarity and control radiation levels during the defueling operation.

The defueling operation at TMI-2 requires that high water quality be maintained. This is necessary to ensure adequate water quality for direct viewing of the defueling operations and to provide that radiation dose rates to workers from suspended solids are as low as reasonably achievable (ALARA). In an undamaged reactor suspended corrosion products can cause a visibility problem. The TMI-2 reactor sustained damage creating a significant quantity of

0097-6156/86/0293-0239$06.00/0
© 1986 American Chemical Society

fine debris which can become suspended when disturbed during the
defueling activities. Therefore, maintaining the water clarity
within the TMI-2 reactor is a challenging engineering problem.

Two objectives of adequate water clarity and reduced worker
dose rates can be maintained if the turbidity of the borated water
is ≤ 1 NTU. The Defueling Water Cleanup System (DWCS) has been
designed by GPU Nuclear Design Engineering (DE) to achieve these
objectives. The primary component in the DWCS is the Filter
Canister.

Babcock & Wilcox (B&W) under contract to GPU Nuclear Corporation has developed a containerization system consisting of three
canisters. One of these canisters is the Filter Canister. The
Filter Canister is a component in both the DWCS and the Fines/Debris
Vacuum System (F/DVS). The Filter Canister is designed to remove
suspended debris particles from reactor coolant water. Externally,
the filter is similar to other canister types. These are described
in detail in reference (1).

There are several filter performance criteria important to
B&W's effort in the development of a filter canister. These are as
follows:

1. The filter should be capable of removing fine debris sized above
   0.5 micron to a maximum particle size of 800 micron.
2. The effluent turbidity from the filter should be equal to or
   less than 1.0 NTU.
3. The Filter Canister's operating life should be sufficiently long
   to allow the canister to accumulate 400 pounds of debris.
4. The Filter Canister should have a minimum flow capacity of 100
   gallons per minute.
5. The chemical nature of the borated water should not be altered
   by the filtering process.
6. The Filter Canister configuration should provide a critically
   safe geometry for a 30-year storage period.
7. The overall filtration system should have a straightforward
   operational format that minimizes complexity and promotes system
   reliability.

Background

Based on both the Quick-Look inspection and core debris sample
analysis, it has been determined that the TMI-2 reactor contains
significant quantities of debris fines with a particle size of 40
microns or less. In addition, other fines are expected to be generated a a result of the defueling operation.

The debris fines less than 40 microns in size will remain suspended indefinately in the water surrounding the reactor core.
These suspended debris fines contribute to the anticipated high turbidity in the reactor core and must be removed. To remove the fines
and reduce the turbidity, a filtration system capable of maintaining
water clarity at less than 1 NTU is required.

Early in 1983, GPUN considered a variety of filtration systems.
Alternates reviewed included etched disc filters, sintered metal
filters, and ultrafiltration. The evaluation precluded the use of
organic polymer-based filters because of the extended radiation
exposure that the filter canister would receive. The early filter

studies were reported in a paper presented at the Waste Management
'84 Conference (2).   As a result of that work, GPUN identified a
backwashable tubular porous 0.5-micron sintered stainless steel
filter concept for the TMI-2 water treatment system.

The filter offered the potential of a relatively high flow rate
in a self-contained unit. By being backwashable, the solids removal
capacity of the filter could be expected to be high. In addition, a
nominal rating of 0.5 micron suggested the filter could effectively
remove a high percentage of the submicron particles in the reactor
coolant.

Two filter vendors who manufactured a 0.5-micron porous
sintered 316L stainless steel filter media were Mott Metallurgical
Corporation and the Pall Trinity Corporation.   Preliminary filter
feasibility tests were conducted by both vendors.

As a result of the preliminary vendor tests, GPUN decided to
pursue the Mott filter concept.   Proof-of-principle testing of the
Mott filter was initiated by B&W.   During the Mott filter testing
program the Pall Trinity Corporation provided a different sintered
stainless steel filter concept.   The new Pall concept offered the
potential of having significant advantages over the tubular filter
element and it was incorporated into B&W's filter test program.

Filter Concepts

Mott Metallurgical Filter Concept.   The Mott filter concept is based
on a porous sintered 316L stainless steel tubular filter element
(0.75-inch OD/0.625-inch ID/6-foot length).   Solids are loaded on
the inside diameter of the filter at a constant flowrate until a
predetermined differential pressure across the filter is reached.
The system then automatically shuts down and the filter is hydraulically backpulsed to remove the collected solids. After allowing the
solids to settle into a plenum area, the filter operation is restarted.   The filter is operated in a cyclic mode until the plenum
is loaded with solids.

A conceptual Mott filter canister is depicted in Figure 1. B&W
conducted extensive testing of a single Mott filter element. Proof-
of-principle testing was demonstrated at this scale.   Further
full-scale testing of the Mott concept was suspended when the Pall
filter concept was identified. Reasons for adopting the Pall Filter
Concept include:

1. Elimination of backwashing.
2. Continuous filtration operations versus cyclic operation.
3. Elimination of a complex filter control system.

Pall Filter Concept.   The Pall Trinity Micro Corporation produced a
new filter media which is a combination of woven stainless steel
wire mesh with stainless steel powder sinter bonded to its process
stream surface.   The media is approximately 0.016-inch thick and can
be fabricated into pleated cartridge type module.

The cylindrical filter cartridges are 11 inches long with a
2.5-inch OD (Figure 2).   In operation, the filter cartridge is
loaded with solids on the outer diameter surface at a constant flowrate until a differential pressure of 40 psid is reached.   At 40
psid the filter cartridge will have collected sufficient solids to
consider the canister fully loaded.

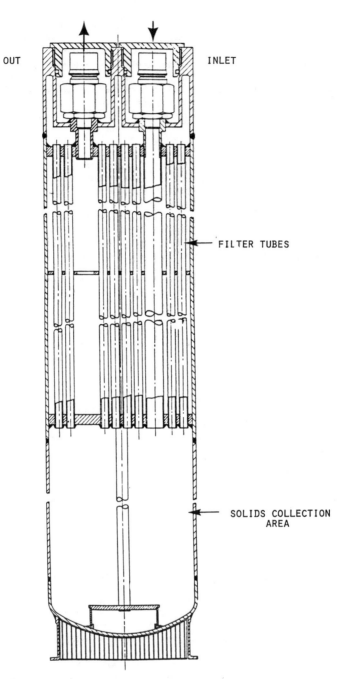

OUT

INLET

FILTER TUBES

SOLIDS COLLECTION
AREA

Figure 1.   Conceptual Mott Filter Canister.

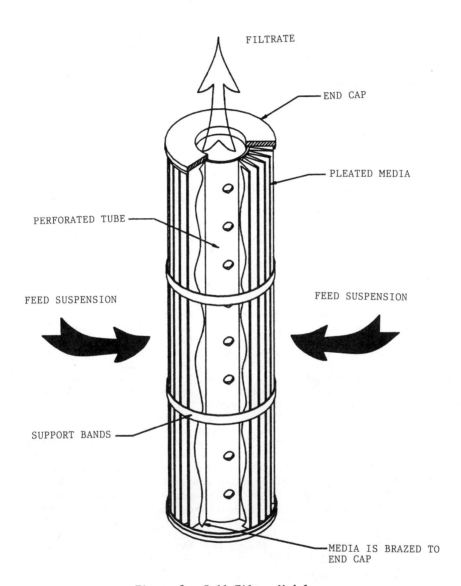

Figure 2. Pall Filter Module.

The Pall filter module is the basic building block of the filter canister. Eleven filter modules are welded end-to-end to form a filter element.

The internal drain tube of the element is plugged at the top. The drain tube at the lower end is sealed into a header and support plate. The finished filter canister consists of 17 filter elements in a concentric circular pattern (Figure 3). The Pall filter canister has a surface area of 523 square feet.

Filter Testing

The filter tests were conducted in an automated state-of-the-art filter test facility designed by B&W engineers. A block diagram of the facility is provided in Figure 4.

The filter-test-facility is designed to provide a candidate filter cartridge with a constant flow rate of liquid containing a fixed concentration of suspended solids. The suspended solids concentration can be changed by adjusting the flow from a slurry injection system prior to the main feed pump or by adjusting the concentration in the slurry tank. Solids-free liquid is supplied from a second, larger tank which is constantly refilled by the recycle filtration system from the effluent hold-up tank. This arrangement allows testing of filters at almost any solids concentration and total process volume. In-line instrumentation monitors effluent turbidity, differential pressure, and flow rate. A process-controller is included in the control system for automated operation as well as a computer controlled data collection system.

The solid fines used in the tests were chosen to simulate the actual particle concentration, size, and density as close as possible without using radioactive materials. The only exception was the use of uranium dioxide in several tests. The composition of the solid fines used for testing was based on actual fuel debris analysis and predicted defueling system performance.

The filter candidates were challenged with simulated debris solids suspensions of 1400 and 140 ppm in a borated water solution (5000 ppm boron). The solids fines used included zirconium dioxide, stainless steel, iron oxide (red) and for several tests uranium dioxide was used. Test data generated included solids loading, effluent turbidity, and differential pressure as a function of time.

The filter test parameters used are presented in Table 1. The solids fines particle size distribution information is presented in Table II.

Table I. Filter Test Parameters
Feed Solution: pH adjusted 5,000 ppm borated water

Flow Density (GPM/Sq Ft)	Mott	0.5		
	Pall	0.25		
	DWCS		**F/VDS**	
Solids Concentration (ppm)	140		1400	
Solids Composition (w/o)	ZrO_2	85	SS/UO_2	60
	Fe_2O_3	15	ZrO_2	30
			Fe_2O_3	10

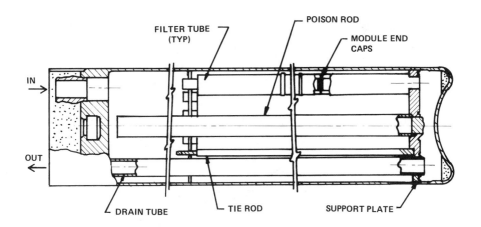

FILTER CANISTER — CROSS-SECTION AT MID-PLANE

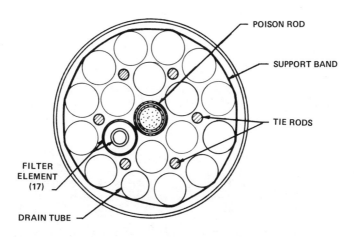

Figure 3.   Filter Canister - Pall Design.

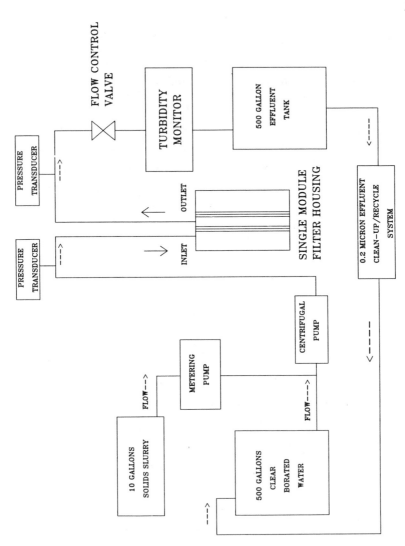

Figure 4. B&W Filter Test System.

Table II.   Solid Fines Test Materials

Zirconium Dioxide	
Zirox-70	95% < 2.1 micron
	40% < 0.5 micron
Zirox-250	95% < 20 micron
	40% < 1.6 micron
Iron Oxide	
Fe_2O_3 (red)	95% < 1.3 micron
Uranium Oxide	
UO_2	95% < 8 micron
	40% < 1 micron
Stainless Steel	
316L powder	95% < 42 micron
	40% < 16 micron

Test Results

The Mott filter concept was evaluated in a series of 14 tests. These represent a total test period of 300 hours. The operation of the Mott single filter element was successfully demonstrated.   The filter's effluent turbidity was less than 1.0 NTU. A final filter canister loading of 400 pounds could be projected. The turbidity and differential pressure trends for a typical Mott filter test are illustrated in Figure 5.

The Pall filter concept was evaluated in a series of 12 tests, representing a total of 250 hours of testing. The operation of a single filter module was successfully demonstrated. The filter's effluent turbidity was maintained less than 1.0 NTU. The Pall Filter Canister can be projected to remove more than 400 pounds of debris. The differential pressure trends for a Pall Filter module test are given in Figure 6.

A summary of the test results for each filter at 1400 and 140 ppm is given in Table III.

Table III.   Summary of Test Results

	Mott Filter Element Test Results		
	Flowrate: 0.5 GPM/Sq Ft, 0.56 GPM		
Solids Conc. (ppm)	Cycle Time (min)	Turbidity (NTU)	Solids Loading (lbs)
1400	71	<1.0	4.2
140	342	<1.0	1.6
	Pall Filter Module Test Results		
	Flowrate: 0.25 GPM/Sq Ft		
Solids Conc. (ppm)	Cycle Time (min)	Turbidity (NTU)	Solids Loading (lbs)
1400	503	<1.0	4.1
140	1755	<1.0	1.4

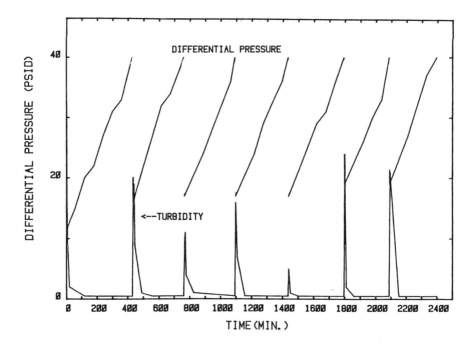

Figure 5.   Differential Pressure and Turbidity vs. Time (Mott Element).

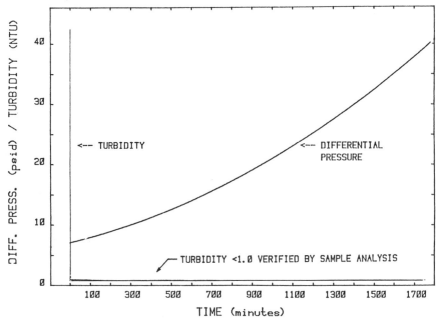

Figure 6.   Differential Pressure and Turbidity (Pall Module).

Conclusions

The Pall Filter Concept offered significant advantages over the Mott Filter Concept. The Pall Filter was chosen for the final filter design based on its simplified continuous mode of operation.

The full-scale Filter Canister is designed to remove suspended solids from 800 micron to 0.5 micron. The filter Canisters are scheduled to be in service the summer of 1985.

Literature Cited

1. P.C. Childress and E.J. McGuinn. "TMI-2 Core Removal Program Defueling Canister Design," Waste Management '85, 1985.
2. K.B. Rao and W.H. Bell. "Developments in Backwashable Fine Filter," Waste Management '84, 1984.

RECEIVED July 16, 1985

14

Cleanup of Demineralizer Resins

W. D. Bond[1], L. J. King[1], J. B. Knauer[1], K. J. Hofstetter[2], and J. D. Thompson[3]

[1]Oak Ridge National Laboratory, Oak Ridge, TN 37831
[2]GPU Nuclear Corporation, Middletown, PA 17057
[3]EG&G Idaho, Inc./TMI, Middletown, PA 17057

Radiocesium is being removed from demineralizers A and B (DA and DB) by a process that was developed from laboratory tests on small samples of resin from the demineralizers. The process was designed to elute the radiocesium from the demineralizer resins and then to resorb it onto the zeolite ion exchangers contained in the Submerged Demineralizer System (SDS). It was also required to limit the maximum cesium activities in the resin eluates (SDS feeds) so that the radiation field surrounding the pipelines would not be excessive. The process consisted of 17 stages of batch elution. In the initial stage, the resin was contacted with 0.18 $\underline{M}$ boric acid. Subsequent stages subjected the resin to increasing concentrations of sodium in NaH_2BO_3-H_3BO_3 solution (total boron = 0.35 $\underline{M}$) and then 1 $\underline{M}$ sodium hydroxide in the final stages. Results on the performance of the process in the cleanup of the demineralizers at TMI-2 are compared with those obtained from laboratory tests with small samples of the DA and DB resins. To date, 15 stages of batch elution have been completed on the demineralizers at TMI-2, which resulted in the removal of about 750 Ci of radiocesium from DA and about 3300 Ci from DB.

As a consequence of the accident at the Three Mile Island Nuclear Power Station, Unit 2 (TMI-2) on March 28, 1979, the two demineralizers (DA and DB) in the water makeup and purification system were severely contaminated with fission product radionuclides. The resin beds in the demineralizers were significantly degraded both radiolytically and thermally by the decay of most of the radionuclides during the period since the accident. The principal gamma-emitting radionuclides remaining on the resin beds when demineralizer cleanup activities began were due to the relatively long-lived ^{137}Cs ($t_{1/2}$ = 30.1 y) and ^{134}Cs ($t_{1/2}$ = 2.06 y). Prior to the inception of the present investigation, nondestructive assay (NDA) methods had been employed to estimate the cesium activity and material content of

0097-6156/86/0293-0250$06.00/0

each demineralizer (Table I) (1). Comparison of the post-accident
with pre-accident resin bed volumes indicated that the beds had
undergone severe shrinkage (~55%) and significant degradation had
occurred. It was known that the beds had not only been subjected to
high radiation dosages (~10⁹ rad) but had also been exposed to high
temperatures because of the radioactive decay heat. The necessity
to isolate the demineralizers from liquid flow about 19 h after the
accident prevented effective removal of the decay heat, and estimates indicate that center-line bed temperatures may have been as
high as about 540°C (1000°F). The demineralizers were sampled by
GPU Nuclear personnel in early 1983, and it was observed that the DA
vessel contained only dry, caked, resin, whereas liquid was still
present in DB. However, the caked bed (DA) was apparently deagglom-
erated after water addition and sparging so that resin samples were
obtained in a later sampling effort. The absence of liquid in the
DA vessel is contrary to the NDA estimate of 3 ft³ (85 L).

Table I.   Estimated Demineralizer Vessel Loadings
Based on NDA Characterizations[a]

| | | Post-accident | |
Loadings	Pre-accident[b]	DA	DB
Resin			
Volume, ft³	50	22	22
Weight, lb	2,139	1,025	1,025
^{137}Cs, Ci	0	3,500	7,000
^{134}Cs, Ci	0	270	540
Liquid			
Volume, ft³	44	3	3
Weight, lb	2,746	193	193
Debris			
U, lb		5	1
Core debris, lb		95	19
^{137}Cs, Ci		177	35
^{134}Cs, Ci		16	3
^{106}Ru, Ci		21	4
^{144}Ce, Ci		28	5
^{125}Sb, Ci		116	23
TRU, Ci		0.5[c]	0.1[c]

[a]Data obtained from M. K. Mahaffey et al. (1).
[b]DA and DB vessel loadings are identical.
[c]Alpha activity only.

Conceptual studies of the various alternative methods for
cleanup of the demineralizers (2) indicated that the most desirable
method was to elute the cesium and subsequently sorb it on the
zeolites in the Submerged Demineralizer System (SDS) (3-6). This
concept alleviated the high-level radiation problems associated with
eventual removal of the degraded resin beds and with management of
the resin wastes. Since the effects of the resin bed degradation
were unknown with regard to the quantitative elution behavior of
cesium and the quality of the eluates, this concept required

experimental determination of its feasibility and the development of a satisfactory chemical flowsheet. A program was established in early 1983 involving the collaborative efforts of Oak Ridge National Laboratory (ORNL), General Public Utilities Nuclear Corporation (GPU Nuclear), and EG&G/Idaho, Inc./TMI (EG&G) to: (1) establish the technical feasibility of cesium elution, (2) develop a chemical flowsheet that would meet processing requirements for the SDS at TMI-2, and (3) provide for cleanup of the demineralizers. ORNL was responsible for establishing the feasibility of cesium elution and for developing the chemical flowsheet using demineralizer samples provided by GPU Nuclear. GPU Nuclear, with some assistance from EG&G, was responsible for installation of the process at TMI-2 and subsequent cleanup of the demineralizers.

Several requirements had to be met in the development of a satisfactory process flowsheet for cesium elution and fixation. In elution, it was necessary that chemical reagents be employed that were compatible with the ionic solution chemistry of SDS feed solutions. This requirement dictated that elution of the cesium be accomplished by displacement with sodium ions using sodium borate or sodium hydroxide solutions. The elution process was also required to limit the cesium activity in eluates to levels that were no greater than about 1 mCi/mL to avoid excessive radiation fields in subsequent SDS operations. Even with compatible ionic solution chemistry, it was by no means certain that eluates could be satisfactorily processed in the SDS. Cesium loading of the SDS zeolite beds might be seriously impaired by the presence of degradation products in eluates, depending on their quantity and nature. Finely dispersed solids (or colloids) from resin particle breakage and degradation might not be readily separable and therefore could cause plugging of the zeolite bed. Finally, soluble organic compounds and/or emulsified oils might sorb on zeolites and foul or block the exchange sites for cesium.

Laboratory Development Studies

Laboratory-scale experiments on the development of the process flowsheet included: (1) batch elution tests to determine the feasibility of cesium elution by sodium ion displacement; (2) tests with small zeolite beds to evaluate the feasibility of satisfactorily processing eluates in the SDS; and (3) studies of eluate clarification by settling and by filtration through sintered-metal filters. The results of these experiments provided the technical basis for the process flowsheet for the cleanup of the demineralizers. The DB sample was available about 1 year before a satisfactory DA sample was obtained; therefore, most of the elution flowsheet conditions were developed using the DB resin.

<u>Description and Analysis of Demineralizer Samples.</u> Samples from the DA and DB vessels consisted of resin mixed with liquid. The DA sample contained about ~20 mL of resin and ~50 mL of liquid, and the DB sample contained about 40 mL of resin and 80 mL of liquid. The liquids were separated from the resins by settling and decantation; most of the resin settled very rapidly. The separated resin was then dried in air at ambient conditions (25 to 30°C). In some

cases, the separated liquid was further clarified by centrifugation
before analysis. The resin and liquid were analyzed by chemical and
radiochemical methods. The resin was also examined by visual
microscopy at ~20X magnification to assess its physical characteristics.

Radionuclide and chemical analyses of the resin and liquid
phases of the DA and DB samples are given in Table II (only the
principal constituents are listed). Small quantities (10 to 1000
ppm) of many other metal cations were also present in the resin, but
they are not relevant to the work reported here.

Table II. Radionuclide and Chemical Analyses of the
Demineralizer Samples

	DA		DB	
Analyses	Liquid[a]	Resin	Liquid[b]	Resin
Radionuclides, $\mu Ci/g$				
^{137}Cs	209	5,520	1,480	21,800
^{134}Cs	11	285	101	1,458
^{90}Sr	6.72	3,060	9.46	890
Chemical, ppm				
C	164	--	950	--
B	1,000	20	2,000	>200
Na	500	4,900	8,500	>1,000
SO_4[c]	900	29,000	7,700	15,000
U	<1	2,400	10	200
Fe	4	2,400	10	200
Ca	3	970	15	30
Ba	-	240	--	<1

[a]pH = 7.1.
[b]pH = 5.7.
[c]Sulfur was determined as sulfate.

Microscopic examinations revealed that the DA resin was more
severely damaged than the DB resin (Figure 1). The DA resin contained larger angular particles and clustered resin beads in
significant amounts, perhaps 5 to 10 vol % of the sample. However,
the remainder of the particles were clearly distinguishable as individual resin beads, with colors ranging from black (nontransparent)
to amber (transparent). The angular particles had an appearance
that is typical of pyrolytic carbons derived from tars, pitches, and
polymeric resins.

The angular particles were not observed in the DB resin, which
principally contained only black or amber-colored beads; very few
bead clusters were observed. In some cases, partial spalling of the
surface layer of the black beads occurred, exposing an amber-
colored, transparent interior. The blackening only appeared to
occur to a depth of a few micrometers.

Liquids separated from the demineralizer samples were visually
clear but yellowish brown in color. It was later shown in liquid
clarification tests that the solutions were slightly turbid even
after settling for 1 to 4 d.

DA Resin

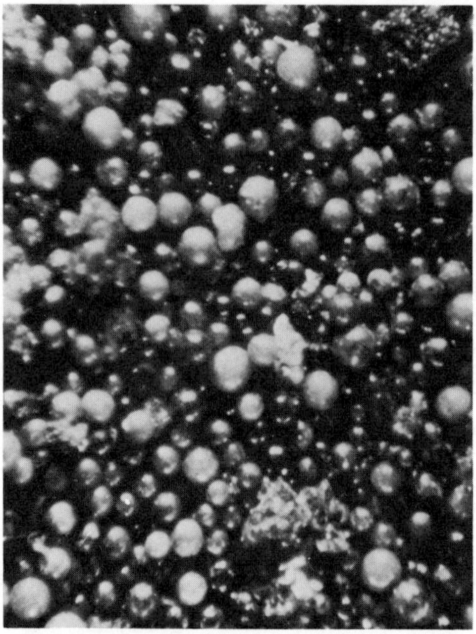

DB Resin

Figure 1. Photomicrographs of the Demineralizer Resins
(Magnification, 20X).

Multistage Batch Elution Tests. The results from multistage batch tests that demonstrated a satisfactory method for cesium elution are shown in Table III and Figure 2. The elution process consisted of 17 contact stages in which a sodium-ion concentration gradient was employed to limit the cesium activities of the eluates to a maximum of ~ 1000 $\mu Ci/mL$ (Table III). The resin was initially rinsed with 0.18 $\underline{M}$ H_3BO_3 in the first stage, and then the sodium-ion concentration of the eluent was increased in succeeding elution stages from 0.035 to 1 M through the use of partially neutralized boric acid solutions ($\overline{Na}H_2BO_3-H_3BO_3$) and, finally, 1 $\underline{M}$ NaOH.

After 17 stages, 96% of the ^{137}Cs had been eluted from the DB resin, but only about 56% had been eluted from the more degraded DA resin (Figure 2). The specific activities of the ^{137}Cs remaining on the DA and DB resins were 2.4 and 0.9 mCi/g, respectively. Partial elution of the ^{90}Sr also occurred, and about 25% was eluted from either resin. The distribution coefficient (K_d) values of ^{137}Cs for the DB resin were relatively constant throughout the stage contacts with the same sodium concentration, whereas a significant increase in K_d values for the DA resin occurred (Table III). This difference in elution characteristics of ^{137}Cs can be attributed to the greater damage suffered by the DA resin.

The eluted resin bed was air-dried, classified into three fractions by screening, and analyzed to determine whether the large, clustered, and carbonized particles in the DA resin sample contained most of the uneluted [137]Cs. Only about 12 to 14 particles were present in the >2000-μm fraction, about 10 vol % of the bed was in the 850- to 2000-μm fraction, and about 90 vol % was in the <850-μm fraction. The specific activities of ^{137}Cs in the resin were 4.1, 4.7, and 2.4 mCi/g for the >2000-, 850- to 2000-, and <850-μm fractions, respectively. Although the larger particles are higher in specific activity, most of the uneluted [137]Cs ($\sim 80\%$) is associated with the <850-μm fraction.

Organic compounds were solubilized throughout the 17 stages of batch contacts (Figure 3), but only to the extent of about 8 mg of carbon per gram of resin. Eluates from the elution of the DA and DB resins showed no significant difference in dissolved organics. No attempt was made to identify these compounds.

The clarity of eluates appeared to be generally good by visual observations. However, some cloudiness was observed in the first stages of elution with the DB resin and during the final treatment with 1 $\underline{M}$ NaOH with either the DA or DB resin. The eluates from stages $\overline{1}$ and 4 of the DB resin elution were filtered through a 0.5-μm-rated nylon filter disk, and the filter was subsequently scanned by gamma spectroscopy (Table IV). Only small quantities of radioactivity were detected, principally due to ^{137}Cs and much lesser amounts of [125]Sb, [144]Ce, and [60]Co. The observed activity of ^{137}Cs was about 0.01% of that of the filtered solution (~ 400 MBq).

Zeolite Bed Tests. Decontamination factors (DFs) for ^{137}Cs and ^{90}Sr were determined in the zeolite bed tests with resin eluate feeds and with the liquid separated from the as-received DA sample that had been diluted 20:1 with 0.18 $\underline{M}$ H_3BO_3 (DA liquid). The tests consisted of passing 70 to 100 bed volumes of feed through a 2-mL bed of mixed zeolites (60 vol % Ionsiv IE-96 and 40 vol % Linde A-51).

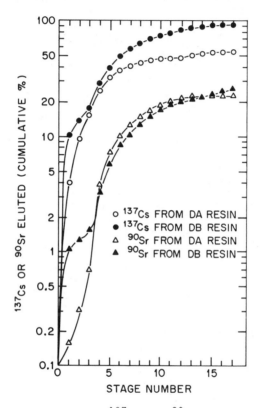

Figure 2.   Percentage of ^{137}Cs and ^{90}Sr Eluted in Multistage Batch Tests.

Table III. Elution of ^{137}Cs in Multistage Batch Elution Tests
with DA and DB Resins

Conditions: 15 mL of eluent solution contacted with ~10 mL
of resin for 30 min; air-dried resin weights
were 6.56 and 7.22 g for the DA and DB resin,
respectively.

Stage No.	Eluent	^{137}Cs Conc. in Eluate (μCi/mL) DA	DB[a]	K_d, $\frac{\mu Ci/g}{\mu Ci/mL}$ DA	DB[a]
1	0.14 M H$_3$BO$_3$	96.8	1090	54.8	17.9
2	0.035 M NaH$_2$BO$_3$—0.32 M H$_3$BO$_3$	136	340	36.7	55.4
3		135	420	34.7	42.8
4	0.26 M NaH$_2$BO$_3$—0.09 M H$_3$BO$_3$	245	1230	16.8	12.5
5		185	1140	20.0	11.4
6		122	1050	28.0	10.3
7		78.1	870	41.5	10.5
8		67.3	757	45.9	10.0
9		51.9	632	57.2	9.9
10		37.0	516	78.0	10.1
11		25.5	405	111	10.8
12		18.1	343	154	10.9
13	1 M NaOH	64.3	432	41.0	6.6
14		42.7	395	59.5	5.1
15		25.7	264	96.5	5.6
16		16.0	175	153	6.4
17		21.1	97	200	9.7

[a]Eluent volumes in stages 7, 12, and 17 were 14, 13, and 13 mL,
respectively.

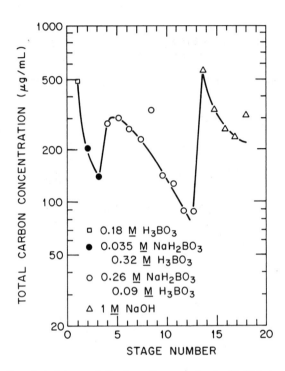

Figure 3. Leaching of Carbon Compounds in Multistage Batch
Elution of the DB Resin.

Table IV.   Gamma Scan of Filter Solids
from a DB Resin Eluate

Radionuclide	Activity (Bq)
^{60}Co	27
^{125}Sb	310
^{134}Cs	1.66×10^3
^{137}Cs	2.69×10^4
^{144}Ce	52

The superficial bed residence time was kept constant at 8.4 min
throughout each test, which is a value comparable to that ($\sim$10 min)
used in SDS processing at TMI-2.   Test results are shown in Table V.
    The cumulative DFs for ^{137}Cs range from about 5 x 10^3 to 3 x
10^5; those for ^{90}Sr range from about 200 to 500.   The DF values and
their variation with the feed concentrations of ^{137}Cs and ^{90}Sr are
consistent with results of previous studies (3,4) on the performance
of the SDS zeolites.   The ^{137}Cs DF for the DA liquid feed was
slightly improved by filtering it with a 10-μm-rated filter frit,
perhaps indicating that a very small percentage of the ^{137}Cs (0.01%)
in the feed was associated with the nonsorbable and finely dispersed
solids that were present.   No evidence of bed plugging was noted in
any of the tests, even though the feeds for Runs Z-2, Z-3, and Z-4
were not filtered.
    Some sorption of the soluble organic compounds in the eluate
feeds on the zeolite bed occurred.   Cumulative breakthroughs for the
organic compounds for the DA and DB resin were 45 and 75%, respectively.   However, there was no significant change in the DF for
^{137}Cs, while the bed loading of organics increased in a near-linear
fashion during the tests.

Liquid Clarification.   Clarification of the DA liquid was investigated by two methods: filtration and settling.   Filtration tests
were conducted with sintered-metal (stainless steel) filter disks
with ratings of 0.5 and 10 μm at a filtration rate of 2.2 mL per cm^2
per min.   This flow rate was the design basis for the filtration
system to be used at TMI-2, which also provided for periodic
backflushing when the filter pressure drop reached 20 psi.   The
liquid used for the filtration and settling tests was separated from
the DA sample after 8 h of settling of the initially well stirred
sample and was subsequently diluted 20-to-1 with 0.18 $\underline{M}$ boric acid
(2000 ppm boron).
    The results of the filtration test (Table VI) showed that good
clarification was obtained using either the 0.5- or the 10-μm-rated
filters.   Turbidity was decreased from 12.6 to about 2 nephelometric
turbidity units (NTUs) using either filter.   The 0.5-μm-rated filter
rapidly plugged (16 to 19 min), whereas the 10-μm-rated filter took
170 min to plug.   Onstream filtration times of 170 min were judged
marginal for the clarification operations at TMI-2 with backflushing; therefore, a 20-μm-rated filter was installed at TMI-2 as
an additional precaution.   Also, it remained to be demonstrated in
actual operations at TMI-2 that the filter can be satisfactorily
restored by backflushing.   The volume of sample available was too
small to permit investigation of filter backflushing or of extensive
studies of filtration.

Table V. Cumulative Decontamination Factors for ^{137}Cs and ^{90}Sr in Zeolite Bed Tests

Conditions: 2-mL mixed zeolite bed (60 vol % Ionsiv 96– 40 vol % Linde A-51); residence time = 8.4 min

Test	Source	Feed Solutions			Effluent bed volumes	Cumulative DF	
		^{137}Cs (µCi/mL)	^{90}Sr (µCi/mL)	Total C (µg/mL)		^{137}Cs	^{90}Sr
Z-2	DB resin eluates;[a] stages 1-13	346	0.14	80	80	1.50 E5[b]	5.11 E2
Z-3	DA resin eluates;[a] stages 1-13	97	32.5	130	70	>3.48 E5	4.96 E2
Z-4	DA liquid	10	0.26	8	100	4.34 E3	2.60 E2
Z-5	DA liquid; filtered through 10-µm-rated filter	8	0.18	12	100	1.99 E4	2.03 E2

[a]From elutions described in Table III.
[b]Read as 1.50 x 10^5.

Table VI. Results of Clarification Tests with Sintered-Metal
Disk Filters using DA Liquid

Conditions: 9.5-mm-diam filter disk, filtration rate =
2.2 mL/cm^2·min; DA liquid diluted 20:1
with 0.18 $\underline{M}$ H$_3$BO$_3$

Run No.	Filter rating (μm)	Time (min)	Final pressure drop (psi)	Volume (mL)	Turbidity[a] (NTUs)
1	0.5	19	20	30	2.1
2	0.5	16	20	25	2.3
3	10	170	14[b]	270	2.2

[a]Turbidity of feed was 12.6 NTUs.
[b]Feed supply was exhausted. Extrapolation of the pressure-drop-vs-
time curve indicated that 20 psi would be reached after 190 min and
a filtrate volume of ~300 mL.

Settling tests (Table VII) showed that clarification could be
significantly improved by using longer settling times. Therefore,
relatively long settling times after batch contacts of resin and
eluents could be employed to reduce filter loadings in eluate
clarifications and to increase liquid throughputs before excessive
filter plugging occurred. Relatively long settling times (24 to
72 h) were used in the processing of the eluates at TMI-2 (discussed
later).

Table VII. Effect of Settling Time on
DA Liquid Turbidity

Settling Time (d)	Turbidity (NTUs)
0	12.6
1	6.1
2	5.5
3	4.6
4	4.0

An attempt was made to increase the settling rate by using
fresh anion exchange resin (Amberlite IRA-400; 20-40 mesh) as a
flocculation aid in one test; however, it did not significantly
improve the rate unless prohibitively large quantities were added.
In that experiment, 1-g additions of resin were made each day to the
same 35-mL aliquot of liquid until a total of 4 g had been accumulated. The resulting turbidities of the liquid after settling for
1 d in the presence of 1, 2, 3, and 4 g of the resin were 5.2, 4.4,
3.5, and 2.8 NTUs, respectively.

Process Flowsheet Description and Process Operations at TMI-2

The schematic process flowsheet is shown in Figure 4. After the
existing liquid in the demineralizers had been removed, cesium was
eluted in multistage batch contacts of the resin with the eluent
solutions. As elution proceeded, the sodium concentration of the

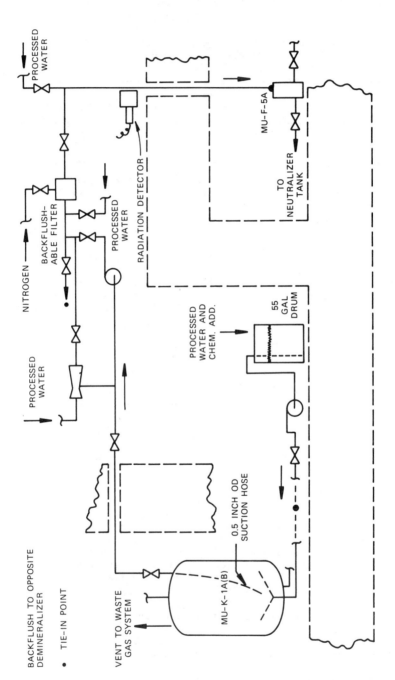

Figure 4. Flowsheet for Elution of the TMI-2 Makeup and
Purification Demineralizers.

eluent solution was increased from 0 to 1 $\underline{M}$ (0 to 23,000 ppm sodium)
to facilitate elution of the residual cesium from the resin. A
liquid/solid phase volumetric ratio of ~1.5 was used in each stage
and is governed by the available free volume in the demineralizer
vessels. Batch contacts of the liquid eluent solutions with the
resin was accomplished by upflowing the liquid into the demineralizer and then air sparging the liquid above the resin bed to promote mixing for 8 h. The air sparging was then stopped, and 24 to
72 h was allowed for the suspended solids to settle. Typically,
three to four drums (55 gal each) of eluent solution were added per
batch contact. Eluates were withdrawn through a suction hose from
about 1 ft below the liquid surface to prevent the withdrawal of any
solids which might be floating on the liquid. The removal was
accomplished by using either an eductor or a pump to lift the
liquid. The eluates were then filtered through a sintered,
stainless steel filter frit (pore size, 20-μm rating) before being
diluted with processed water (800 ppm boron), transferred to in-
plant neutralizer tank storage, and finally processed through the
existing SDS (3-6). The final filtration polish of the eluates
avoids the unnecessary introduction of fine particulate matter to
the SDS and subsequent water management systems.

Eluate dilution was necessary to minimize personnel exposure
during transfer through piping to in-plant tanks. Most of the
eluates were transferred from the demineralizer using the eductor
that dilutes the eluates (up to 20:1) before filtration. Flow
totalizers were used to measure volume of liquid transfer, and the
flow of eluate from the demineralizer continued until the pump or
eductor loses suction, leaving a considerable residual volume (V_r)
of eluate heel in the demineralizer. The values of V_r remaining in
the DA and DB vessels after transfers were 145 and 175 gal, respectively.

Samples from the neutralizer tank were analyzed after each
batch elution. Routine analyses include determination of ^{137}Cs by
gamma spectrometry, ^{90}Sr by beta-counting, and sodium by atomic
absorption. The progress of the batch elution was monitored from
the determination of the concentration and volumes in the neutralizer tank before and after transfer of batch eluates.

Processing Results. After 15 stages of batch elution, approximately
750 and 3300 Ci of ^{137}Cs have been eluted from DA and DB, respectively (Table VIII). Process operations have been generally satisfactory, but some difficulty was encountered due to frequent filter
plugging when eluting DB with sodium concentrations greater than
about 0.25 $\underline{M}$ (8000 ppm). Filter plugging was minimized and satisfactory filtration rates were obtained by increasing the settling
time of the eluates from the nominal 24 h to 72 h and by the
addition of boric acid to reduce solution alkalinity. It is
believed that filter plugging was due to degraded resin product particles that were more effectively dispersed at the highly alkaline
conditions; this phenomenon was not observed in the DA elution.
Results from the processing of the filtered eluates through the SDS
system have been completely satisfactory.

Based on the NDA estimates of the ^{137}Cs demineralizer loadings
in Table I, about 22 and 46% of the ^{137}Cs have been eluted from DA

Table VIII. Cesium Removals Accomplished after 15 Stages of
Batch Elution of the TMI-2 Makeup and
Purification Demineralizers

	Demineralizer A		Demineralizer B	
Stage No.	Cumulative ^{137}Cs removal (Ci)	Sodium conc. of eluent ($\underline{M}$)	Cumulative ^{137}Cs removal (Ci)	Sodium conc. of eluent ($\underline{M}$)
1	119	0[a]	811	0[a]
2	201	0[a]	932	0[a]
3	206	0.07	1190	0.07
4	395	0.34	1510	0.07
5	464	0[b]	1940	0.34
6	499	0.23	2000	0.34
7	532	0[b]	2180	0[b]
8	562	0.23	2420	0.30
9	608	0.74	2640	0.30
10	630	0.38	2840	0.23
11	646	0[b]	2920	0.30
12	706	0.38	3070	0.51
13	711	0.90	3170	0.51
14	745	1.0	3240	0.59
15	745	1.0	3260	0.59

[a]Processed water containing 0.18 $\underline{M}$ H_3BO_3.
[b]Removal of filter backflush liquid.

and DB, respectively (Figure 5). The removal of ^{137}Cs was about
one-half of that expected on the basis of the hot-cell tests at
ORNL, if it is assumed that the samples tested were representative
of the bulk resin from each demineralizer resin bed. The liquid-
resin contact was definitely better in the laboratory hot-cell tests
because the resin was slurried with the liquid via shaking.
Nevertheless, the trends in ^{137}Cs elution behavior with increasing
sodium concentration were expected to be the same. The in-plant
data from the DB resin elution show the expected trend; however, the
DA data do not, particularly at the higher sodium concentrations
approaching 1 $\underline{M}$. The results to date indicate that little additional ^{137}Cs can be removed by continuing the DA elutions with 1 $\underline{M}$
sodium concentrations. Thus, it appears that either the original
cesium loading estimates for DA (Table I) are high or the resin
sample used in the hot-cell elution tests is not representative of
DA resin bed.
     Resin sampling procedures with the demineralizers using quartz
fiber optical examination of the DA resin bed showed that it was
dry, which is contrary to the wet-resin-bed model used to interpret
the NDA data. The possibility that the original estimate of the
[137]Cs loading on DA is high is currently being explored in calculations using a revised source term model to interpret the NDA data.
No attempts at further elution of DA are planned, pending the out-
come of the reassessment of the ^{137}Cs loading estimate and radiation
level surveys of the DA vessel. However, elutions will continue for
DB at higher sodium concentrations to further reduce the cesium
activity. Following completion of the cesium elution operations,

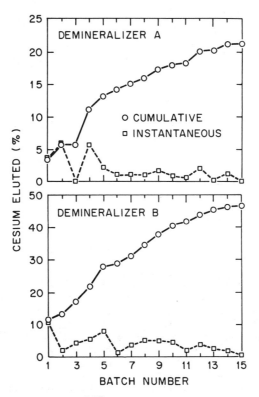

Figure 5.   Percentage ^{137}Cs Eluted in the Multistage Batch Elution of the TMI-2 Demineralizers.

dose-rate surveys of the demineralizer cubicles will be performed to
confirm the final resin elution efficiencies and the dose rate
reductions that were achieved.

Acknowledgments

The authors wish to express their sincere appreciation to several
individuals at ORNL and GPU Nuclear for valuable assistance in this
work. Gratitude is expressed for the excellent services provided by
the ORNL Analytical Chemistry Division personnel under the direc-
tions of J. A. Carter, J. H. Cooper, D. A. Costanzo, and J. M.
Peele. Special thanks are due the ORNL Metals and Ceramics Division
personnel under the supervision of R. Lines, who provided the photomicrographs of the demineralizer resins. Sincere appreciation is
also acknowledged to GPU personnel under the direction of B. G.
Smith for capably performing the demineralizer cleanup operations at
TMI-2 and the associated analytical work.

Literature Cited

1. M. K. Mahaffey et al. "Resin and Debris Removal System
    Conceptual Design--Three Mile Island Nuclear Station Unit 2
    Makeup and Purification Demineralizer," Hanford Engineering
    Development Laboratory Report HEDL-7335, Richland, Washington,
    May 1983.
2. E. J. Renkey and W. W. Jenkins, "Planning Study of Resin and
    Debris Removal Study, Three Mile Island Nuclear Power Station
    Unit 2 Make-up and Purification Demineralizers," Hanford
    Engineering Development Laboratory Report HEDL-7377, Richland,
    Washington, June 1983.
3. D. O. Campbell, E. D. Collins, L. J. King, and J. B. Knauer,
    Evaluation of the Submerged Demineralizer System (SDS) Flowsheet
    for Decontamination of High-Activity-Level Water at the Three
    Mile Island Unit-2 Nuclear Power Station, Oak Ridge National
    Laboratory Report ORNL/TM-7448, Oak Ridge, Tennessee, July 1980.
4. D. O. Campbell, E. D. Collins, L. J. King, and J. B. Knauer,
    "Process Improvement Studies for the Submerged Demineralizer
    System (SDS) at the Three Mile Island Nuclear Power Station,
    Unit 2," Oak Ridge National Laboratory Report ORNL/TM-7756,
    Oak Ridge, Tennessee, May 1982.
5. L. J. King, D. O. Campbell, E. D. Collins, J. B. Knauer, and
    R. M. Wallace, "Evaluation of Zeolite Mixtures for
    Decontamination of High-Activity-Level Water in the Submerged
    Demineralizer System (SDS) Flowsheet at the Three Mile Island
    Nuclear Power Station, Unit 2," in Proceedings of the Sixth
    Internatinal Zeolite Conference, Reno, Nevada July 10-15, 1983,
    D. Olson and A. Bisio (Eds.), Butterworths, 1984, pp. 660-68.
6. K. J. Hofstetter, C. G. Hitz, T. D. Lookabill, and S. J.
    Eichfield, "Submerged Demineralizer System Design, Operation,
    and Results," Proc. Topl. Mtg. Decontamination of Nuclear
    Facilities, Niagara Falls, Canada, September 19-22, 1982, Vol.
    2, pp. 5-81, Canadian Nuclear Association (1982).

RECEIVED July 16, 1985

The Recovery: A Status Report

John C. DeVine, Jr.

GPU Nuclear Corporation, Middletown, PA 17057

This paper provides an up-to-date synopsis of the Three Mile Island Unit 2 (TMI-2) Recovery Program. The discussion is presented within the context of the three-phase program approach that is followed by the GPU Nuclear/Bechtel Group integrated organization, in restoring safe stable conditions at TMI-2, following the March 1979 accident.

In the first few weeks after the accident, efforts centered upon regaining control and assessing the damage incurred. The early recovery work centered around the new systems required to supplement or replace damaged systems and components, and the new procedures needed to deal with the damaged plant.

This technical summary describes the subsequent work including Phase II, Defueling, and Phase III, Cleanup, leading to future decisions and work efforts. To date, the TMI-2 recovery program has been successful in achieving:

o A high degree of public and worker safety

o Progress in the recovery program, resulting in
   a vast improvement at TMI-2

Moreover, the program has proved to be a significant learning experience for the entire nuclear industry, which is enhancing the safety of nuclear plant operations.

0097-6156/86/0293-0267$06.00/0
© 1986 American Chemical Society

More than six years have passed since the morning of March 28,
1979, when the now-famous accident at Three Mile Island Unit 2
(TMI-2) suddenly transformed the subject of nuclear safety from
scientific hypothesis to reality. In that time, the TMI-2 accident
and cleanup have become perhaps the most-reported and
least-understood events of our time.

Today, the prevailing public and political notion of the TMI-2
cleanup seems to be that it is a very hazardous venture, plagued
with problems and achieving little, if any, success. Much press
coverage of the cleanup has been negative. Within the technical
community as well, frustration is often expressed about the
seemingly slow pace of the cleanup work.

For these reasons, we welcome this opportunity to provide an
up-to-date, factual synopsis of our TMI-2 Recovery Program. In the
next few pages I will summarize what we have accomplished, where we
stand, and what lies ahead in this important program.

First, three points of primary importance must be made:

1. The TMI-2 accident and recovery should be (and to an
    extent, have been) the most significant learning
    experience for our industry. Undoubtedly, every
    nuclear plant is safer now than it would have been had
    the accident not happened. And much more is to come.
    For example, we are just now beginning to accumulate
    hard information from the damaged core itself, to
    serve as a basis for invaluable research on
    temperatures achieved, heat transfer and chemical
    processes in the core during the accident, fission
    product retention, and the like.
2. The cleanup has been conducted with absolutely highest
    regard for worker and public safety, and it has been
    extremely successful on that score. Despite
    allegations and insinuations to the contrary, that
    record speaks for itself. Offsite doses have been
    negligible, in-plant personnel exposures have been
    well controlled and are at very low values,
    individually and collectively.
3. The cleanup program, although hardly problem free, has
    achieved real progress toward its ultimate goal. In
    every respect, the plant condition is now vastly
    improved over that facing the team at the program
    outset.

With that as background, I will now describe the Recovery
Program.

The Starting Point. In the first few weeks after the TMI-2
accident (of March 28, 1979), an around-the-clock, emergency
response mode of operation prevailed. During this period positive
control was regained, but the situation which emerged--and became
the starting point for the recovery effort that is still in
process--was discouraging indeed.

Although the reactor's condition was not precisely known, it
was clear that massive fuel damage had occurred. The reactor
containment building was completely inaccessible because of high
radiation and radioactive contamination levels, and most of the
Auxiliary and Fuel Handling Building (AFHB) complex was
inaccessible or marginally accessible as well. Hundreds of
thousands of gallons of highly contaminated water had collected in
the reactor building basement and in Auxiliary Building tanks.
Because of the high contamination levels (ranging from one to more
than 100 microcuries per mililiter gross activity), plant systems
were inadequate to transfer or process this water.

These technical difficulties were compounded by such problems
as continuing public fear, regulatory uncertainty, funding
limitations, and closure of commercial radwaste disposal sites (for
TMI-2 wastes).

Early Recovery Work. Work began immediately on a variety of
parallel programs to address these problems. Numerous new plant
systems (including several decay heat removal systems, a new
nuclear sampling system, a major RCS pressure and inventory control
system, and numerous instrumentation and control systems) were
designed, procured, and installed on an urgent schedule to replace
or supplement existing systems not considered suitable for extended
use because of inaccessibility, radiation levels, and the like.
New procedures were developed to deal with the plant in its damaged
condition. Design work began on two major liquid waste processing
systems, and facilities were constructed to safely stage solid and
liquid wastes for extended periods of time. Aggressive
decontamination of the AFHB was conducted, and preparations were
initiated for re-entry into the containment building.

As work proceeded during the first few years of the recovery
program, organizational and institutional steps were taken as
well. The present organization--which combines elements of GPU
Nuclear and the Bechtel Group into a unique, integrated unit
structured to deal with long-term recovery--was formed in 1982.
Project funding from multiple sources, along the lines initially
proposed by Governor Thornburg of Pennsylvania, was secured.
Cooperative agreements among the U.S. Department of Energy (DOE),
the U.S. Nuclear Regulatory Commission (NRC), and GPU Nuclear
greatly facilitated the recovery effort by providing DOE (and
National Laboratory) support for various program activities of
national interest and permitting research and disposal of TMI-2
radioactive wastes in U.S. Government facilities. Agreements among
GPU Nuclear, DOE, and a group of Japanese government, utility, and
nuclear industry companies resulted in Japanese participation and
funding for research-related aspects of the program.

Also during the period, strategies and plans for the entire
recovery effort began to take shape. The concept of a phased
project, as depicted graphically in Figure 1, was established and
continues in use today.

Per this concept, the recovery program consists of three
distinct (but overlapping) phases. The objective of Phase I (which
is complete) was to regain and maintain stable plant conditions for

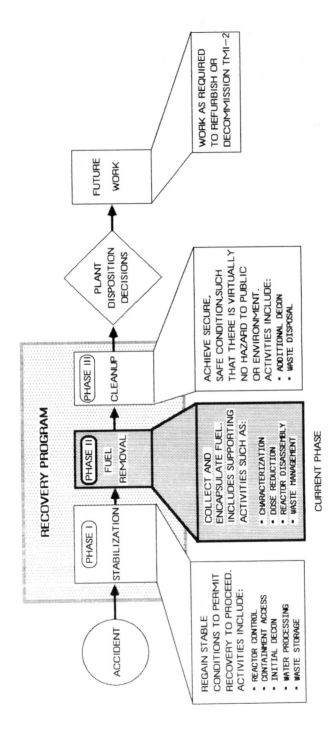

Figure 1. TMI-2 recovery program phases.

the extended term. This phase consisted primarily of those early
activities outlined above. The objective of Phase II (the current
phase) is to remove and encapsulate the damaged fuel from the
reactor, a major project in its own right. Phase III then will
eliminate residual radiation hazards, resulting in a plant
condition that is safe and secure. A decision as to the ultimate
plant disposition, which could range from recommissioning to any of
several decommissioning phases, will not be made until after Phase
III. At this point, all safety issues will have been resolved and
plant conditions will permit its thorough examination to support
such a decision. Of course, TMI-2 work required after that point
will depend upon that determination.

Phase I is history, and therefore needs no further discussion.
The following is a technical summary of Phases II and III, and the
work to be done after their completion.

Phase II - Defueling. Within the overall TMI-2 Recovery Program,
reactor defueling is the most visible, expensive, and technically
challenging element. Preparations for the defueling effort--in the
form of conceptual plans, data acquisition, engineering,
procurement, plant refurbishment, equipment installation, and
training--have been underway for about three years. These
preparations are nearly complete.

One key task of the defueling project is data acquisition and
analysis to characterize the conditions of fuel and structural
material inside the reactor. Because of high radiation and
contamination levels, this work must be done remotely.

A variety of techniques have been used in this effort,
including small diameter TV camera examination inside the reactor,
removal and analyses of fuel and structural material samples,
internal and external ion chamber measurements, sonar profiling of
the core void region, external profiling of neutron flux using
solid state track readers (SSTRs), and others. These examinations
and their results are being described in some detail in other
presentations. In summary, however, two points are important:

> From an engineering standpoint, a thorough
> understanding of conditions inside the reactor vessel
> is a major prerequisite to reactor disassembly and
> defueling, both for safety reasons and to ensure that
> the chosen defueling techniques are adequate for their
> intended applications. This examination program has
> also produced information of great value to the
> scientific community in analyzing the TMI-2 accident.
> The examinations have been conducted in a very
> difficult and unique environment, and have involved
> the development and/or refinement of innovative
> techniques, equipment, and analytical methods. These
> developments have been very successful and have broad
> application and value to the industry.

At this point, a relatively clear picture of the damaged TMI-2
reactor is emerging. Visual examination (confirmed by its
demonstrated integrity over the six years since the accident) has
shown that the reactor vessel itself is intact and in sound

condition. Similarly, the reactor vessel head and the upper
reactor internals are highly contaminated but relatively damage
free. However, damage to the nuclear fuel is extreme.
    As originally installed, the nuclear core consisted of 177 fuel
rod assemblies, arranged in a cylinder approximately 12 ft high and
12 ft in diameter. Virtually the entire upper five feet of this
core region is now a void, and the three feet below that is a mass
of loose rubble. Samples extracted from this rubble bed confirm
that some of the fuel reached very high temperatures (at or
approaching the melting point of uranium oxide fuel) during the
accident. The lower part of the core region has not been examined
directly but is expected to include a spectrum of conditions; from
intact fuel rods to agglomerated material and voids. Finally, a
substantial amount of core material (perhaps 10-20 tons, assumed to
be fuel and structural material) has been discovered below the core
region and appears to have been once molten.
    Conceptual plans, detailed engineering, and design,
fabrication, and installation work have been proceeding for the
systems and equipment needed to safely remove, encapsulate, store,
and transport this damaged core material. In very brief summary,
the defueling process is as follows:
    1. Several methods will be used to extract fuel from the
        reactor, including vacuum systems, long-handled tools,
        and remote manipulators.
    2. Using these methods under water, the fuel will be
        placed in stainless steel canisters and these will be
        closed securely.
    3. Using specially designed handling equipment, the fuel
        cans will be moved into the TMI-2 fuel handling
        building, and then placed in underwater racks for
        temporary storage in the spent fuel pool.
    4. The fuel cans will then be dewatered, placed in
        specially designed rail casks, and transported to
        Department of Energy facilities in Idaho for research,
        storage, and ultimate disposal.
    Several major reactor disassembly steps (including polar crane
repair, reactor head removal, and upper plenum jacking) have
already been completed in preparation for the defueling effort.
Numerous dose reduction tasks (such as decontamination and
shielding), defueling equipment installation and testing,
preparation of procedures, and personnel training are in progress
and nearing completion.
    The actual defueling work will commence this year, starting
with removal of the loose rubble material in the upper core region
and then proceeding to the more difficult regions below. The
entire job is expected to take about two years to complete.

Phase III and Beyond. After completion of the reactor vessel
defueling, several other significant operations are required to
achieve the safe, secure, and accessible condition that is the
object of the recovery program. These include:

Collection and encapsulation of the particles of fuel
transported to portions of the reactor coolant system
and connected auxiliary systems during the accident.
While the absolute quantity of such fuel is expected
to be relatively small (in aggregate, probably less
than one percent of the core), this will be a
difficult job because of access limitations,
radiological conditions, and the like.
Substantial additional decontamination of the reactor
containment and auxiliary buildings. Earlier
decontamination work in these buildings was (and is)
somewhat selective, intended to reduce radiation and
contamination levels in areas where plant
stabilization and defueling work (i.e., Phases I and
II) required frequent access. The Phase III
decontamination effort has the broader objective of
achieving conditions that are satisfactory for the
longer term in the sense that they pose virtually no
risk of release of radioactive material, and they
permit access as necessary to thoroughly examine the
plant for determination of its ultimate disposition.

A major part of this decontamination effort will involve the
reactor building basement. This area is essentially inaccessible
now because of high radiation levels (ranging from a few R/Hr to
over 1000 R/Hr). Much work has already been done to prepare for
basement decontamination, including data acquisition and
development of robotic equipment.

As we approach Phase III, additional work will be done to
secure and isolate plant systems. Also, various systems and
equipment will be installed to monitor and control the defueled
plant. This Phase III work is, for the most part, still in the
planning stages, although some activities (such as Auxiliary
Building Decon) are proceeding already on a not-to-interfere
basis. A key element in the planning process is the development of
a licensing strategy, tailored to the unique TMI-2 condition, which
will afford the proper balance of practicality and plant
protection. Development of this strategy along with a related set
of specific Phase III end point conditions, is now underway.

After Phase III, a decision can be made as to the ultimate
disposition of the plant. The obvious disposition candidates are
decommissioning (either in the near term or at some point in the
future, such as when TMI-1 is decommissioned), or refurbishment and
recommissioning.

The decision will require extensive study, taking into account
technical economic, and regulatory factors. Such studies,
particularly those requiring assessment of the physical condition
of the plant, cannot be completed until Phase III completion
criteria have been established. Therefore, a decision as to the
ultimate disposition of TMI-2 is not anticipated until 1989, or
later.

Conclusion

This presentation has been necessarily brief. The TMI-2 Recovery
Program is, in a sense, a composite of a number of smaller
projects--many of which are technically significant but could not
be summarized here. Questions on any aspect of the program are
welcome.

In conclusion, let me restate the obvious. The Three Mile
Island accident is an event of extraordinary impact on GPU Nuclear,
on the nuclear industry, on the electric utility industry, and on
our energy future.

There are those who would exploit the TMI-2 experience by
calling it the final, compelling demonstration of the failure of
nuclear power and a deterrent from further nuclear power plant
development.

Nothing could be further from reality. The TMI-2 accident, and
the recovery that has followed, present an opportunity of
inestimable value. It is our challenge to extract from the
experience every possible benefit, to learn from our mistakes and
to apply these lessons to the design, regulation, and operation of
nuclear power plants. On that basis, we can proceed safely,
strongly, and successfully.

RECEIVED July 29, 1985

Impact on Future Licensing

W. F. Pasedag[1] and A. K. Postma

U.S. Nuclear Regulatory Commission, Washington, D.C. 20555

The TMI-2 accident has had a dramatic impact on the
assessment of severe accidents, particularly on accident
source term assumptions. TMI not only demonstrated that
regulatory interest in severe accidents is appropriate,
but also illustrated our limited understanding of fission
product behavior under degraded core conditions. The
resulting reassessment of accident source terms has
resulted in a concerted, world-wide research effort,
which has produced a new source term estimation
methodology. In order to assess the potential impact of
the application of this methodology on regulatory
requirements, a comparison with the approach used in
licensing analyses is necessary. Such a comparison
performed for the TMI-2 accident sequence, shows that
differences in assumptions concerning accident
progression far outweigh the differences in the
methodology per se. In particular, the degree of
conservatism incorporated into assumptions concerning
operator action and containment response has over-riding
influence on source term estimates. A major contribution
to the impact of the new source term methodology on
regulatory requirements, therefore, is its capability to
provide the improved level of understanding necessary for
reassessment of regulatory assumptions in this area.

The TMI-2 accident has had a dramatic impact on severe accident
assessment, particularly in the area of accident source term
assumptions. The events of March 28, 1979 painfully illustrated our
limited understanding of fission product behavior under degraded
core conditions. Comparisons between observed releases of iodine
and regulatory assumptions for design basis accident calculations,
or the fission product transport assumptions of the Reactor Safety

[1] Current address: GPU Nuclear Corporation, Middletown, PA 17057

Study (1) were drawn to illustrate the need for accident source term
reassessment (2). The NRC staff agreed that a re-evaluation of
accident source terms was appropriate (3,4). The ensuing world-wide
research effort has produced a new source term estimation
methodology. Now that results of calculations with this new
methodology are becoming available, it is appropriate to assess the
technical bases for changes in the regulatory requirements and
licensing practice.

Comparison of Regulatory Assumptions with TMI-2 Observations

In order to assess the potential impact of the application of the
emerging source term methodology on regulatory requirements, it is
necessary to define the baseline of the current regulatory
methodology for source term assessment. This is of particular
interest for the fission product iodine, as it pervades the
regulatory requirements related to accidental releases of
radionuclides. The regulatory assumption of a 100% release of the
core's noble gas inventory, 50% of the iodines, and 1% of the solid
fission products (5) is well known. The conditions under which
these assumptions apply, however, are often mis-understood. The
design basis fission product release assumptions of 10CFR100, for
example, frequently are compared with the estimated TMI-2 release of
15 Ci of iodine without recognition of the significant differences
in the accident sequences. Even the more correct comparison of
WASH-1400 source term estimates for a small loss-of-coolant accident
with the observations at TMI-2 is difficult to interpret, as it
involves source term comparisons for two entirely different accident
sequences. Such comparisons include the effects of the accident
sequence as well as that of the methodology, and, therefore, can be
used to characterize differences in the methodology only if it
assumed that all accidents will follow the same sequence of events
as TMI-2.
     In order to separate the effects of different assumptions
concerning the course of the accident from the differences arising
from source term modeling, it is necessary to compare the results of
the regulatory models with the actual observations for exactly the
same sequence of events. A calculation to facilitate such a
comparison of the results of applying the WASH-1400 assumptions
concerning iodine behavior to the TMI-2 accident sequence is given
in Appendix A. This calculation uses WASH-1400 methodology only,
without any recognition of the recent research results obtained
since the accident. The iodine released to the containment, for
example, is assumed to be in the elemental form, with 0.7% of the
release assumed to be converted to organic iodides, as postulated in
the RSS. (In contrast, current research indicates that dominant
iodine form in severe accidents is likely to be cesium iodide, with
only a minor fraction in volatile form). This calculation produces
the surprising result of excellent agreement of the WASH-1400 source
term methodology with the iodine measurement in the post-accident
containment at TMI-2. Although the basic premise of this
calculation, i.e., iodine release in elemental form only, is not
consistent with other data from TMI-2, a small fraction of volatile,
reactive iodine in the containment atmosphere is not inconsistent

with the TMI-2 data. The result obtained suggests that differences
in the methodology may not be the most important reason for the
previously noted differences between the results of regulatory
accident analyses and the TMI-2 observation. Rather, the
significant differences are attributable primarily to differences in
the assumptions concerning the course of events of design basis
accidents, or the risk-dominant core-melt sequences identified in
WASH-1400, and the actual course of events experienced at TMI-2.
The major differences between the assumed and the experienced
accident sequences are readily identified. During the TMI-2
accident, an essential containment function, i.e., retention of the
radionuclides released to the containment atmosphere, was
maintained, while the containment is assumed to leak or fail in the
regulatory analyses used in the comparisons; the TMI-2 fission
products escaped the primary system through water, as opposed to dry
release pathway assumptions; and, most importantly, the TMI-2 event
was eventually terminated by operator action, while the
risk-dominant accident sequences of the Reactor Safety Study (RSS)
invariably proceed through complete core melting and vessel
melt-through.

Characteristics of Source Term Estimates

The exercise described above demonstrates a general characteristic
of accident source term analyses i.e., the results strongly depend
on the exact sequence of events postulated. If the sequence of
events is known (as it is, in hind sight, for the TMI-2 accident)
source term estimates with different models, performed by various
organizations, generally are in surprisingly good agreement (6).
Slight variations in the accident progression, differences in plant
design, or intervention by the operator, however, can produce widely
different results. In situations where the details of the sequence
of events are unknown, or are postulated on the basis of probability
arguments, the thermal hydraulic analysis serves to define such
critical events as the degree of core oxidation, the on-set of
melting, the rate of fuel heat-up, or the quantities of steam,
hydrogen and suspended liquid in the containment.

The most important issue addressed by the thermal hydraulic
analyst is the question of containment failure. It is generally
agreed that the off-site releases are insignificant if containment
failure (or by-pass) is avoided. Further, even for sequences where
eventual failure of the containment is postulated, a delay of that
failure by several hours results in substantial reduction of the
airborne radionuclide inventory available for release from the
containment (6).

Potential Regulatory Impact

The current regulatory structure incorporates the concept of a small
number of "worst case" accidents, which have become codified as
"design basis accidents". This concept includes a single set of
accident fission product release assumptions for all reactors,
regardless of design (i.e., the "TID-14-844 release") (5). This
uniformity is achieved at the expense of realism, i.e.,

non-mechanistic assumptions are substituted for a realistic
assessment of the phenomena affected by specific plant
characteristics. The existing regulatory structure, therefore, does
not readily lend itself to incorporation of realistic, highly
plant-specific source term information. On the other hand, the
results of the new source term methodology are highly plant and
sequence specific. A revision of the regulatory assumptions by
simply adjusting the numerical values of the existing set of source
term assumptions, therefore, would not only be difficult to achieve,
but would not fully reflect the insights gained since the TMI-2
accident.

It is evident, however, that a large number of specific
regulatory assumptions and criteria need to be re-evaluated. The
most important of these is the basis of the present emergency
planning criteria. This subject area is one of the few places in
the current regulatory structure where a complete spectrum of
accident sequences, including accidents outside of the design basis
envelope, are considered. Incorporation of realistic source terms
for the spectrum of postulated accident sequences into the planning
basis, therefore, should not be very difficult.

A somewhat more difficult task is the evaluation of the
appropriate corrections to the numerous applications of the DBA LOCA
source term. The tabulation of licensing guidance documents
directly affected by the DBA LOCA source term, shown in Table I,
illustrates the pervasive presence of the elemental iodine
assumptions in the current regulatory structure. Although the
underlying assumptions are derived from the single (although
non-mechanistic) accident characterized in TID-14844, the resulting
criteria are applied to a variety of accident sequences.
Incorporation of greater realism in source term assumptions,
therefore, will require evaluations for the specific sequence of
events addressed in each document.

A third category of regulatory requirements affected by the
improved understanding of accident characteristics gained since the
TMI-2 event are operational testing and surveillance criteria. The
containment leakage testing requirements of Appendix J of 10CFR 50,
for example, are candidates for revision based on a much improved
understanding of the efficiency of the containment function. On the
one hand current source term analyses indicate the importance of
assuring functional isolation of the containment (6). Parametric
studies of accident source terms for various levels of containment
leakage, on the other hand, have demonstrated that leakages far in
excess of the current design basis can be tolerated without
significant impact on source terms for severe accidents (7,8).

Conclusions

TMI-2 has had a profound impact on reactor regulation. Early
regulatory responses centered on short-term plant and procedure
modifications designed to prevent recurrance of TMI-like events. On
a broader scale, the TMI-2 accident demonstrated that events outside
the design basis can happen, and therefore, presented regulatory
questions concerning the adequacy of the current design basis
envelope. In formulating the responses to questions of this nature,

Table I. Licensing Documents Directly Affected
by the DBA-LOCA Source Term

Topic	Regulatory Guide	Standard Review Plan Section
Offsite radiological consequences	1.3, 1.4, 1.7	15.6.5 A,B,C
Containment sprays	1.3, 1.4, 1.7	6.5.2, 15.6.5.A
Containment recirculation filters	1.3, 1.4, 1.52	6.5.1, 15.6.5A
Auxiliary Building filters	1.52	6.5.1, 9.4.2, 3, 4
Main Steam Isolation Valve Leakage Control	1.3, 1.96	6.7, 15.6.5.D
Standby Gas treatment	1.52	9.4.5, 15.6.5
Ice Condenser	--	6.5.2, 3, 4
Containment leakage	1.3, 1.4	6.2.1, 6.2.6, 6.5.3
Dual containment	1.4	6.5.3
Pressure suppression pool	--	6.5.3
Control Room Habitability systems	--	6.4
Post-Accident environment	1.89, 1.97	6.1.1, 6.1.2, 9.3.2

the need for improved understanding of severe accidents in general, and fission product behavior in particular became apparent. Recognition of the latter need has lead to a world-wide research program designed to better characterize radionuclide releases to the environment following severe accidents (i.e., accident source terms). The full impact of improved characterization of accident source terms on reactor licensing is only beginning to be assessed at this time.

The anticipated impact of incorporation of improved realism in source term assessment is often characterized by pointing out the large differences between licensing analyses of design basis accidents, or the characterization of severe accidents in the Reactor Safety Study (WASH-1400), and the observed releases at TMI-2. Such comparisons, however, involve distinctly different accident sequences, thereby combining the effects of the assumptions concerning accident progression with the effects of the calculational methodology. The comparison of the regulatory methodology with the TMI-2 observations given in this paper demonstrates that the differences arising from the methodology are outweighed by the effects of the assumptions concerning the accident sequence. Recognition of this fact should modify expectations of uniform lowering of source term characterizations in licensing calculations as a result of new source term methodology. A careful examination of the assumptions concerning accident progression contained in various licensing analyses appears to be necessary to fully incorporate the improved understanding of fission product behavior gained since the TMI-2 accident into future licensing considerations.

Literature Cited

1. Reactor Safety Study, "An Assessment of Accident Risks in U.S. Commercial Power Plants", WASH-1400 (NUREG-75/014), USNRC, Washington, D.C., 1975
2. M. Levenson and F. Rahn, "Realistic Estimates of the Consequences of Nuclear Accidents", Nuclear Technology 1981, 56, 99.
3. W.F. Pasedag, et al., "Regulatory Impact of Nuclear Reactor Accident Source Term Assumptions", NUREG-0771, USNRC, Washington, D.C., 1981
4. M. Silberberg, et. al., "Technical Bases for Estimating Fission Product Behavior During LWR Accidents," NUREG-0772, USNRC, Washington, DC, 1981.
5. J. DiNunno, et al., "Calculation of Distance Factors for Power and test Reactor Sites", TID-14844, USAEC, Washington, D.C., 1962
6. Report of the Special Committee on Source Terms, American Nuclear Society, LaGrange Park, IL, 1984
7. K.D. Bergeron, et al., "Applications of the CONTAIN 1.0 Computer Code to the Analysis of Containment Loading Under Severe Accident Conditions", 12th Water Reactor Safety Information Meeting, USNRC, Washington, DC. 1984
8. E.A. Warman, "SWEC Investigation of Severe Accident Source Terms", ANS Executive Conference on Ramifications of the Source Term, Charleston, S.C., 1985

APPENDIX A

CALCULATION OF IODINE RELEASES TO CONTAINMENT USING RSS METHODOLOGY

This appendix presents a comparison between iodine behavior observed in the TMI-2 accident and expectations based on WASH-1400 methodology.

Important Aspects of Accident Sequence

Details of the accident sequence that have a dominant impact on iodine transport and leakage are the following:

(1) The transport pathway was through the pressurizer. The pressurizer tank was full of saturated water (Nuclear Safety Analysis Center) during the period when most of the iodine entered the containment atmosphere (Daniel, Pelletier).

(2) Most of the iodine was carried to the containment atmosphere by steam and hydrogen passing through the reactor coolant drain tank (RCDT) which was initially full of water.

(3) Most core damage occurred during the first 3 hours after reactor trip. The main release of fission products from the overheated fuel occurred when the pilot-operated relief valve (PORV) block valve was closed (Daniel, Pelletier, Rogovin).

(4) The PORV block valve was opened intermittently between 3 and 10 hours (Nuclear Safety Analysis Center). Sprays operated for approximately 6 minutes (Rogovin).

(5) At approximately 10 hours following turbine trip, a hydrogen burn pressurized the containment atmosphere and actuated the containment sprays (Nuclear Safety Analysis Center). Sprays operated for approximately 6 minutes (Rogovin).

(6) Average pressure in the containment building was ~2 psig during the period from 3 to 11 hours; the reactor containment building (RCB) was at a negative pressure compared with the outside atmosphere for longer periods (Nuclear Safety Analysis Center).

Iodine Release to the Containment Atmosphere

During the period from 3 to 10 hours, roughly 2.9×10^5 lb of steam and 340 lb moles of H_2 were estimated to have been vented from the pressurizer (Daniel) and to have carried with them approximately 1.8×10^6 Ci of I-131 (Daniel). This represents 2.6% of the core inventory of I-131.

Water in the quench tank was heated to the boiling point by steam from the pressurizer during the first few hours after reactor trip (Daniel). Therefore, for the period when iodine was released to the RCB atmosphere, little condensation would be expected in the RCDT. Trapping of iodine in the RCDT was estimated using the following approximations:

(1) The gas volume passing through the tank was equal to the steam
     and hydrogen volumes discharged from the pressurizer, adjusted
     for thermal conditions in the RCDT.

(2) Iodine in the gas leaving the RCDT was in equilibrium with
     iodine in the liquid.

(3) The iodine partition coefficient was 2.3 x 10^4, based on
     WASH-1400.

(4) The drain tank contained 1220 ft^3 of water (Daniel).

     On the basis of these assumptions, it was estimated that 88% of
I_2 released from the pressurizer would be trapped in the RCDT, and
12% (0.31% of core inventory) would enter the containment
atmosphere. A similar calculation was made for noble gases, and the
predicted retention in the quench tank was less than 0.1%, a
negligible retention compared with that of iodine.
     The 12% of the iodine not trapped in the quench tank was
assumed to enter the containment atmosphere. For the purpose of
this calculation, the iodine entering the containment atmosphere was
assumed to be in the form of I_2 and the source rate was taken as
constant over a 7-hour period starting at 3 hours and ending at 10
hours.

Predicted I_2 Transport in the Containment Building

Plateout of iodine on surfaces, according to the WASH-1400 model
(Ritzman), can be expressed as (Knudsen)

$$\frac{C_g V}{W_0} = \frac{1}{\lambda t_r} (1 - e^{-\lambda t}) \tag{1}$$

where C_g = airborne iodine concentration, Ci/ft^3
             V = volume of contained gases, ft^3
             W_0 = quantity of I_2 released to containment, Ci
             λ = plateout rate constant, hr^{-1}
             t_r = time duration of release, hr
             t = time

     The left-hand side of Equation (1) is the fraction of the total
release that is airborne.
     The value of λ was calculated from

$$\lambda = k_g \frac{A}{V} \tag{2}$$

where λ = plateout rate constant, hr^{-1}
             k_g = mass transfer coefficient, ft/hr
             $\frac{A}{V}$ = surface/volume ratio, ft^{-1}

     k_g was assigned a value of 9 ft/hr based on an estimated film
temperature difference of 2°F (Knudsen), and A/V was assigned a
value of 0.116 ft^{-1} (Daniel).
     Equation (1) was applied for 7 hours starting at 3 hours.

Spray washout of I_2 according to the WASH-1400 model (Ritzman) was estimated to have a rate constant of 0.58 min^{-1} (Stratton). Washout for 6 minutes, starting at 10 hours, was computed using

$$C_g = C_{go} \exp [- 0.58 \ t] \qquad\qquad (3)$$

where C_g = airborne concentration
C_{go} = airborne concentration at t = 10 hr
t = time from the beginning of spray operation, min

After washout and plateout have reduced the airborne I_2 concentration by roughly two orders of magnitude, a pseudoequilibrium was set up. Long-term I_2 re-evolution behavior was predicted using sump volumes estimated from Daniel and Pelletier and partition coefficients given in WASH-1400 (Ritzman).

The airborne iodine concentration predicted on the basis of the models and assumptions described above was calculated for the first 75 hours. The calculated iodine concentration is shown in Fig. 1.

The airborne iodine inventory is predicted to peak at 0.042%. Spray operation at 10 hours rapidly depletes I_2.

Organic Iodide Formation

Organic iodides typically are observed to exist in containment atmospheres following iodine released from fuels or from experiments that release I_2 (WASH-1233). In WASH-1400, a realistic estimate of organic iodide formation is 0.7% when sprays do not operate and 0.4% when sprays operate. The fractional conversions apply to the quantity of iodine that is released to the containment atmosphere. In TMI-2, sprays did not operate except for a brief period; therefore, the predicted formation of organic iodides is (0.007)(0.12)(0.026) = 2.18 x 10^{-5} of the core inventory. This quantity of iodine represents organic iodides and other penetrating gaseous species.

Spray washout of organic iodides by caustic sprays is known to be a slow process and was neglected.

The organic iodine concentration resulting from these assumptions is shown in Fig. 1.

Comparison With Measurements

The first sample of the post-accident containment atmosphere was withdrawn from the containment at 75 hours after reactor trip. This sample indicated that the total airborne I-131 amounted to a fraction of 2.7 x 10^{-5} of the core inventory (Daniel). This sample was drawn through a small-bore steel tube having a length of hundreds of feet, and only penetrating forms of iodine would be expected to be transported efficiently through the sample delivery line. Therefore, it is important to note, the sample would not be expected to verify iodine present as I_2. The predicted quantities, within the containment, are given below.

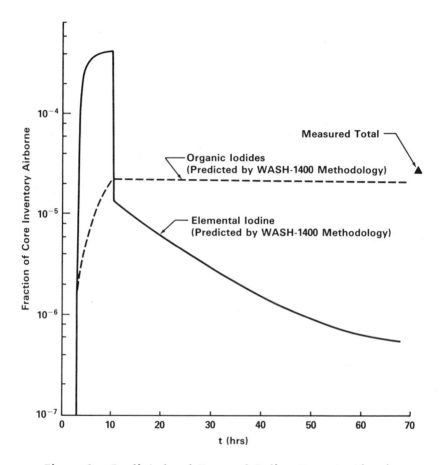

Figure 1.   Predicted and Measured Iodine Concentration in
            Containment Atmosphere

Fraction of Core Inventory Airborne at 75 Hours
Predicted by

Iodine Form	WASH-1400 Methods	Measured
I_2	0.044×10^{-5}	--
CH_3I	2.16×10^{-5}	--
Total	2.20×10^{-5}	2.7×10^{-5}

The predicted and measured quantities of airborne iodine at 75 hours are nearly identical. While the degree of agreement must be regarded as fortuitous, it is clear that the observed iodine concentration is not inconsistent with predictions based on WASH-1400 methodology.

Although the gas sample withdrawn at 75 hours did not discriminate among iodide forms, the sample method (passing of the sample through a long tube) ensured that most of the collected iodine would have been penetrating species. Also, subsequent samples of the containment atmosphere showed that most of the airborne iodine was present as organic iodides (Pelletier). For these reasons, an estimate of organic iodides is just the total iodine in the 75-hour sample: 2.7×10^{-5} of the core inventory.

Response of Dome Radiation Monitor to Spray Actuation

The RCB dome radiation monitor responded to fission products released to the containment atmosphere, indicating that most fission products entered the containment atmosphere between 3 and 10 hours following reactor trip (Pelletier). When the sprays were actuated at ~10 hours and operated for 6 minutes, the radiation levels indicated by dome monitors dropped by a factor of ~30 (Pelletier). This drop in radiation level is consistent with the washout of an airborne species having a high-energy gamma emission.

If it is assumed that the airborne species is I_2, WASH-1400 spray models indicate that a reduction by a factor of ~32 would be expected because of spray operation, which is in excellent agreement with the observation.

Conclusions Related to Transport in the Reactor Containment Buiding

On the basis of the examination of iodine transport in the RCB during the TMI-2 accident with models based on WASH-1400 methodology, the following conclusions could be reached:

(1) The limited data available concerning iodine behavior in the containment atmosphere is not inconsistent with the WASH-1400 methodology, including the assumption of a significant fraction of the iodine in elemental form.

(2) On the basis of WASH-1400 models, most of the iodine released from the core during the initial 10 hours of the accident (88%) was predicted to have been retained in the quench tank.

(3) Of the iodine released to the containment atmosphere, approximately 0.9% is calculated to be organic iodides. This fractional conversion agrees with WASH-1400 assumptions.

LITERATURE CITED IN APPENDIX

J.A. Daniel, et al., "Preliminary Radioiodine Mass Balance Within
the Containment of TMI-2", Report prepared for EG&G Idaho, Inc. by
Science Applications, Inc., February 1982.

J. Flaherty, "Pathways for Transport of Radioactive Material
Following the TMI-2 Accident", Report TDR-055, GPU Nuclear Corp.,
TMI, Middletown, PA, July 8, 1981.

J.G Knudsen and R.L. Hilliard, "Fission Product Transport by Natural
Processes in Containment Vessels: BNWL-943, Battelle Pacific
Northwest Laboratories, Richland, Washington, January 1969.

D.A. Nitti, Babcock and Wilcox Co., Lynchburg, Virginia, Private
Communication with A.K. Postma, March 17, 1982.

Nuclear Regulatory Commission, "Investigation Into the March 28,
1979 Three Mile Accident by Office of Inspection and Enforcement:
NUREG-0600, USNRC, Washington, D.C., August 1979.

Nuclear Safety Analysis Center, "Supplement to Analysis of Three
Mile Island-Unit 2 Accident" NSAC-1 Supplement, Electric Power
Research Institute, Palo Alto, California, October 1979.

C.A. Pelletier, et al., "Iodine-131 Behavior During the TMI-2
Accident" NSAC-30, Electric Power Research Institute, Palo Alto,
California, September 1981.

A.K. Postma and R.W. Zavadoski, "Review of Organic Iodide Formation
Under Accident Conditions in Water--Cooled Reactors" WASH-1233, U.S.
Atomic Energy Commission, Washington, D.C., 1972.

A.K. Postma, "Iodine Evolution from Spray Water" BCT-LR-101, Benton
City Technology, Benton City, Washington, October 26, 1978.

R.L. Ritzman, et al., "Appendix VII to Reactor Safety Study"
WASH-1400, U.S. Atomic Energy Commission, Washington, D.C., August
1974.

M. Rogovin, "Three Mile Island, A Report to the Commissioners and to
the Public" U.S Nuclear Regulatory Commission Special Inquiry
Group. Available from U.S. Nuclear Regulatory Commission,
Washington, D.C.

W.R. Stratton, et al., "Technical Staff Report on Alternative Event
Sequences" Report to President's Commission on the Accident at Three
Mile Island, Washington, D.C., October 1979.

RECEIVED July 31, 1985

INDEXES

Author Index

Subject Index

288

Production by Hilary Kanter
Indexing by Susan Robinson
Jacket design by Pamela Lewis

Elements typeset by Hot Type Ltd., Washington, D.C.
Printed and bound by Maple Press Co., York, Pa.

RECENT ACS BOOKS

"Environmental Applications of Chemometrics"
Edited by Joseph J. Breen and Philip E. Robinson
ACS SYMPOSIUM SERIES 292; 286 pp.; ISBN 0-8412-0945-6

"Desorption Mass Spectrometry: Are SIMS and FAB the Same?"
Edited by Philip A. Lyon
ACS SYMPOSIUM SERIES 291; 248 pp.; ISBN 0-8412-0942-1

"Catalyst Characterization Science: Surface and Solid State Chemistry"
Edited by Marvin L. Deviney and John L. Gland
ACS SYMPOSIUM SERIES 288; 616 pp.; ISBN 0-8412-0937-5

"Polymer Wear and Its Control"
Edited by Lieng-Huang Lee
ACS SYMPOSIUM SERIES 287; 421 PP.; ISBN 0-8412-0932-4

"Ring-Opening Polymerization:
Kinetics, Mechanisms, and Synthesis"
Edited by James E. McGrath
ACS SYMPOSIUM SERIES 286; 398 pp.; ISBN 0-8412-0926-X

"Trace Residue Analysis: Chemometric Estimations
of Sampling, Amount, and Error"
Edited by David A. Kurtz
ACS SYMPOSIUM SERIES 284; 284 pp.; ISBN 0-8412-0925-1

"Polycyclic Hydrocarbons and Carcinogenesis"
Edited by Ronald G. Harvey
ACS SYMPOSIUM SERIES 283; 406 pp.; ISBN 0-8412-0924-3

"Reactive Oligomers"
Edited by Frank W. Harris and Harry J. Spinelli
ACS SYMPOSIUM SERIES 282; 261 pp.; ISBN 0-8412-0922-7

"Reverse Osmosis and Ultrafiltration"
Edited by S. Sourirajan and Takeshi Matsuura
ACS SYMPOSIUM SERIES 281; 508 pp.; ISBN 0-8412-0921-9

"Polymer Stabilization and Degradation"
Edited by Peter P. Klemchuk
ACS SYMPOSIUM SERIES 280; 446 pp.; ISBN 0-8412-0916-2

"Rubber-Modified Thermoset Resins"
Edited by Keith Riew and J. K. Gillham
ADVANCES IN CHEMISTRY SERIES 208; 370 pp.; ISBN 0-8412-0828-X

"The Chemistry of Solid Wood"
Edited by Roger M. Rowell
ADVANCES IN CHEMISTRY SERIES 207; 588 pp.; ISBN 0-8412-0796-8